THE CELL SURFACE
Immunological and Chemical Approaches

ADVANCES IN EXPERIMENTAL MEDICINE AND BIOLOGY

Recent Volumes in this Series

Volume 40
METAL IONS IN BIOLOGICAL SYSTEMS: Studies of Some Biochemical and
Environmental Problems
Edited by Sanat K. Dahr • 1973

Volume 41A
PURINE METABOLISM IN MAN: Enzymes and Metabolic Pathways
Edited by O. Sperling, A. De Vries, and J. B. Wyngaarden • 1974

Volume 41B
PURINE METABOLISM IN MAN: Biochemistry and Pharmacology of Uric Acid Metabolism
Edited by O. Sperling, A. De Vries, and J. B. Wyngaarden • 1974

Volume 42
IMMOBILIZED BIOCHEMICALS AND AFFINITY CHROMATOGRAPHY
Edited by R. B. Dunlap • 1974

Volume 43
ARTERIAL MESENCHYME AND ARTERIOSCLEROSIS
Edited by William D. Wagner and Thomas B. Clarkson • 1974

Volume 44
CONTROL OF GENE EXPRESSION
Edited by Alexander Kohn and Adam Shatkay • 1974

Volume 45
THE IMMUNOGLOBULIN A SYSTEM
Edited by Jiri Mestecky and Alexander R. Lawton • 1974

Volume 46
PARENTERAL NUTRITION IN INFANCY AND CHILDHOOD
Edited by Hans Henning Bode and Joseph B. Warshaw • 1974

Volume 47
CONTROLLED RELEASE OF BIOLOGICALLY ACTIVE AGENTS
Edited by A. C. Tanquary and R. E. Lacey • 1974

Volume 48
PROTEIN–METAL INTERACTIONS
Edited by Mendel Friedman • 1974

Volume 49
NUTRITION AND MALNUTRITION: Identification and Measurement
Edited by Alexander F. Roche and Frank Falkner • 1974

Volume 50
ION-SELECTIVE MICROELECTRODES
Edited by Herbert J. Berman and Normand J. Hebert • 1974

Volume 51
THE CELL SURFACE: Immunological and Chemical Approaches
Edited by Barry D. Kahan and Ralph A. Reisfeld • 1974

THE CELL SURFACE
Immunological and Chemical Approaches

Edited by

Barry D. Kahan
Northwestern University Medical School
Chicago, Illinois

and

Ralph A. Reisfeld
Scripps Clinic and Research Foundation
La Jolla, California

PLENUM PRESS • NEW YORK AND LONDON

Library of Congress Cataloging in Publication Data

Conference Chemical and Immunologic Approaches to the Cell Surface, Augusta, Mich., 1974.
The cell surface: immunological and chemical approaches: [proceedings of an international conference held at Brook Lodge, Augusta, Michigan, June 17-20, 1974]

(Advances in experimental medicine and biology; v. 51)
Includes index.
1. Cell membranes—Congresses. 2. Cytochemistry—Congresses. 3. Immunology—Congresses. I. Kahan, Barry D., ed. II. Reisfeld, Ralph A., ed. III. Title. IV. Series. [DNLM: 1. Cell membrane—Congresses. 2. Cells—Immunology—Congresses. 3. Cytochemistry—Congresses. W1 AD559 v. 51 1974 / QH581.2 C748c 1974]

| QH601.C65 | 1974 | 574.8'75 | 74-20983 |

ISBN-13: 978-1-4684-7244-8   e-ISBN-13: 978-1-4684-7242-4
DOI: 10.1007/978-1-4684-7242-4

Proceedings of an International Conference held at Brook Lodge, Augusta, Michigan, June 17-20, 1974

© 1974 Plenum Press, New York
Softcover reprint of the hardcover 1st edition 1974

A Division of Plenum Publishing Corporation
227 West 17th Street, New York, N.Y. 10011

United Kingdom edition published by Plenum Press, London
A Division of Plenum Publishing Company, Ltd.
4a Lower John Street, London, W1R 3PD, England

PREFACE

The Conference "Chemical and Immunologic Approaches to the Cell Surface" was organized as a multifaceted interaction between scientists representing various disciplines impinging on membrane biology. In many instances, this broad mixture of investigators yielded quite unusual scientific associations and interesting new dimensions to old problems, as the workers came to appreciate the advances, the shortcomings, and the hurdles of each area.

Structural concepts of the membrane--the nature, orientation, and inter-relationship of components--are emerging primarily from work on erythrocytes. Our understanding of surface biology demands reconstruction from the meager, but rapidly emerging, structural information. The excitement of membrane research depends in no small part on the concept that membranes are not static crystalline structures but rather dynamic systems with variable interrelationships between multiple components and phases, reflecting external environmental and internal cellular events. Modulation of the membrane can be readily studied in systems where discrete pertubation is introduced into the surface structure by stimulation with mitogens or reactions of immunoglobulins, resulting in wide-ranging effects. Examination of sequential changes such as patching and capping, in intact cells or in artificially reconstituted lipide, or lipide-protein, membrane systems probably represent useful iatrogenic probes to mimic genuine in vivo biophysical phenomena related to the mobility, cooperatively and constraint of surface components.

The concept of solubilizing membrane components represents a more aggressive approach to the biochemical study of the surface, than does the passive dependence on the analysis of naturally soluble components, such as blood group substances shed into colostrum, ovarian cyst fluid, or gastric mucus. By separating surface glycoproteins, glycolipids, lipids, and protein components from the bulk sea of lipid, solubilization has achieved some success in dissecting the chemical nature of these moieties. Increasingly sophisticated chemical analysis is being used to compare artificially solubilized derivatives with naturally shed materials, such as HL-A antigens, β_2 microglobulin, and immunoglobulins of plasma, in order to understand the relevance of the products to the native state and to thereby reconstruct the interrelationships of macromolecules in the cell surface. The synthesis of such structural and analytic data may yield a better understanding of membrane function.

In many cases, the study of membrane components depends upon biologic assays which represent quite different levels in the pathway from signal reception at the cell surface to target tissue reaction. In hormone research it has been possible to obtain a direct assessment of the precise functional event of reception, namely specific binding, and even measurement of the first stage of translation, such as by detecting activation of adenyl cyclase and membrane enzymes. Release of intramembrane, intracellular, or intercellular, histiotypic messages represents a third level in the hierarchy of biologic assays. Finally, the activity of some surface components can only be detected by effects on homeostatic systems at the level of the intact organism, such as the immune mechanism. In such a case, namely the assessment of histocompatibility and tumor-specific antigens, the chemist must rely upon relatively qualitative, "end-stage" assays, reflecting multiple complex factors remote from the primary specific, membrane interaction.

In choosing this wide domain, the Editors are unabashed by their lack of constraint in the scope of the endeavor. They openly admit to often artificially contriving possible relationships between divergent areas, even when none were immediately apparent. The motivating force for this poetic license was the desire to obtain a wide-ranging discussion of problems, so that experiences of membrane chemists engaged in research in one area might somehow relieve the impasse being encountered in other areas. This endeavor resulted in more than a classification of common problems; the volume apposes information and a critical discussion of data in an original fashion, not even available in the widely divergent, primary source materials.

The Editors are indebted to the American Cancer Society, Merck, Sharpe and Dohme, Hoffman-LaRoche, The Upjohn Company, Meloy Laboratories, Plenum Press, Armour Pharmaceutical, and ICI America for their encouragement and financial support of the conference. Mr. Eugene Wallach and his team performed far beyond their previous heights, by extending from immunology to chemistry, and yet providing an amazingly accurate, rapid transcription of the Proceedings. Ms. R. Jaekel and K. Tamm provided the final solution to the jigsaw puzzle of edited pieces, figures, and direct galley pages.

The prospects for understanding structure and function in the complex, multiphasic membrane system appears bright. The solutions to the riddles of biologic individuality, of immunologic surveillance, and of hormone-target tissue interactions appear almost within our grasp. Hopefully, through this series of Conferences, the premier meeting of which is recorded herein, rapid dissemination of progress in membrane research, and prompt critical analysis by investigators participating in these efforts, will provide a substrate for the development of unified concepts relevant to many disciplines.

CONTENTS

I. Structure and Orientation of Surface Molecules 1
      Session Chairman: V. T. Marchesi

II. Solubilization and Expression of Surface Components 49
      Session Chairman: R. A. Reisfeld

III. Modulation of Cell Surface Structure and Function 101
      Session Chairman: G. M. Edelman

IV. Biological Activities of Solubilized Cell
      Surface Components 163
      Session Chairman: B. D. Kahan

V. Perspective of Cell Surface Structure and Function 243
      Discussant: G. M. Edelman

Conferees 271

Subject Index 277

SESSION I

STRUCTURE AND ORIENTATION OF
SURFACE MOLECULES

Membrane glycoproteins - Glycophorin - Lithium di-iodo-salicylate - SDS-polyacrylamide gel analysis of membrane components - S p e c t r i n - Freeze fracture techniques - Erythrocyte g h o s t s - I n t r a m e m b r a n o u s particles - Ferritin-conjugation electron microscopy - Lymphocyte membrane extrinsic and integral proteins - M i c r o v i l l i - Membrane shedding - Ganglioside structure and function - Cholera toxin.

SESSION I

STRUCTURE AND ORIENTATION OF SURFACE MOLECULES

MARCHESI: Today's session will be divided into three subtopics and each
topic will be introduced by a speaker who will try to give background information and some new experimental data. The first subject will concern itself with the number of different proteins that are present in the red blood
cell membrane. Dr. Heinz Furthmayer will focus on the glycoproteins on
this membrane and present some recent ideas on the orientation of such
molecules. After this subject, Dr. Dan Branton will extend the discussion
to the types of connections between the proteins of the membrane. His recent work has some important implications for the structure of membranes,
and control of receptor sites at the cell surface. Finally, the last of the
major topics will be the third major receptor component on membranes, the
glycolipids. Dr. Pedro Cuatrecasas will discuss some recent work on the
interactions between cholera toxins and gangliosides of membranes, and
this subject will be further amplified by some of the discussants. Then we
will end up the morning session by discussing the question of the mobility
of structures in the membrane, and their relevance to receptor function.

   Before calling on the first speaker I would like to present
Figure 1 which depicts one popular idea concerning the orientation of membrane glycoproteins. As you can see, these molecules may exist in a transmembrane conformation with parts of the polypeptide chain, containing co-
valently bound sugar residues located outside the lipid barrier of the membrane. There is a great deal of complexity in the number and types of sugars
which are on the molecules (Ginsburg will be discussing some of these in-
tricacies tomorrow). The evidence which suggests that this orientation is
more or less correct is rather substantial and I will not go into technical
details now. However, it is important to realize that subtle differences may
exist in the orientation of parts of this molecule which do not obey the general rules which are illustrated in the diagram. Hopefully, some of these
points will come out in the ensuing discussions. It is obvious that the model
proposed above depends upon having a segment of the proteins totally within
the lipid domain of the membrane. One question which has been raised concerning this matter has to do with the type of conformation that the polypeptide would have or might have in the particular circumstance. Many provocative notions have been advanced but it is safe to say that we really have no
indication at all at this time as to what the arrangement really is.

   Another interesting feature of this model which is also slightly controversial is whether the molecule really extends into the cytoplasm
of the intact cell. Again strictly speaking, most of the evidence in support
of this idea is circumstantial and therefore inconclusive. Basically, the
experimental findings in support of this have relied upon the differential

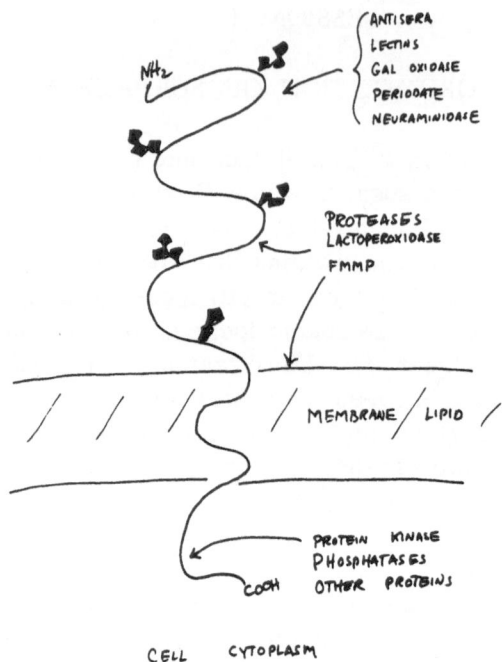

Fig. 1. This is a highly schematic representation of how integral membrane glycoproteins might be oriented at the cell surface.

labelling of parts of the molecule when it is in the intact cell membrane, as opposed to its arrangement in membranes prepared by osmotic lysis. Some critical observers feel that this evidence is weak, and they may be right. However, we will come back to this question later in the discussion, because there is some new experimental data which seem to confirm the model as it was originally proposed. Finally, it is important to state at the outset that if we really want to be critical in terms of evaluating data now at hand, it is important to admit that we do not have any substantive information about this point for any cell type other than the human red blood cell.

Basically, this is all I would like to say as an introduction to this session, although I think it needs to be stressed once again that much of the data relies solely on experimental studies of human red blood cell membranes. For this reason I hope that other discussants will make every effort to bring up data which applies to other cell types when it seems appropriate.

FURTHMAYR: I was asked to give a short introduction to the protein components of the human erythrocyte membrane, and I will do so by showing Figure 2, which demonstrates the SDS-polyacrylamide gel patterns of

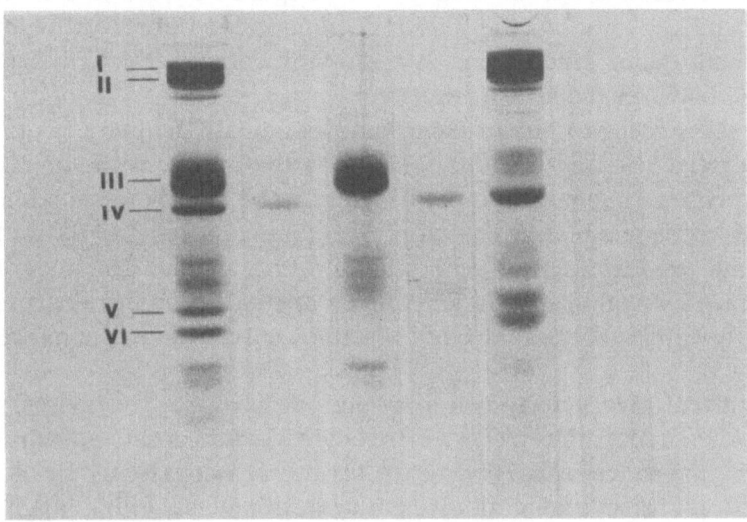

Fig. 2 Membrane proteins analyzed on SDS-polyacrylamide gels. (See text)

various preparations. When membrane proteins are analyzed on SDS-polyacrylamide gels, a number of polypeptides are observed after staining
with Comassie blue, covering a wide range of molecular weights. For
the sake of simplicity and since the function of most of these proteins is
still unknown, numbers 1 - 7 have been assigned to the major polypeptides.
Spectrin is bands 1 and 2; bands 3, 4, 5 and 6 are the predominant species
(on the basis of staining). (Figure 2a) Besides these polypeptides, other
even less well defined components are present in small amounts.

When the gels are stained with periodic acid Schiff reagent for
glycopeptides, usually four bands can be seen: two major bands which are
called PAS_1 and PAS_2, and two minor components, the small molecular
weight component PAS_3 and PAS_4, which migrates between PAS_1 and PAS_2.
(Figure 2b)Today we will certainly hear more about the associations of
these proteins with each other and with other components in the membrane
and about the possible functions of some of these proteins.

These proteins, by virtue of their extractability with salts at high
ionic strength, low ionic strength, and high pH values may be separated
into two major classes. One example is shown in Figure 2 in which case
membranes were extracted with a low concentration of lithium di-iodo-
salicylate (LIS), a procedure which in a single step removes non-inte-
gral proteins from these membranes. After removal of the insoluble
pellet (Figure 2c, d), analysis of the supernate (Figure 2e, f) reveals
the presence of bands 1 and 2 (Spectrin), band 4 which apparently splits
into two bands in low concentrations of SDS in this electrophoretic system,

in addition to bands 5 and 6. The pellet (Figures 2c,d) on the other hand, contains bands 3, 7 and all the major glycopeptides described before. Some overlap might occur since some of the minor components are about equally distributed between these two fractions.

According to our present knowledge, all the proteins found in the LIS extract are lined up in some fashion on the inside of the membrane, while the proteins found in the pellet seem to be tightly associated with lipid. Thin sections of LIS vesicles, the residue after LIS extractions, demonstrate preservation of lipid bilayer structures. The proteins associated with these membraneous structures are called integral proteins, and they usually can be separated from the lipid only by using some kind of detergent.

I will give briefly two examples of how one can isolate such proteins. In the case of the major erythrocyte glycoprotein, advantage was taken of the presence of carbohydrate structures covalently linked to the polypeptide chain, by using an affinity system (Wheat-germ agglutinin-Sepharose). A mixture of total membrane proteins, dissolved in SDS, was applied to the column which had been prepared with these beads in SDS. This column specifically retains only the glycopeptide species PAS_1 and PAS_2, which can be eluted with the specific simple sugar NAcGlc.

Figure 3 shows the result obtained by the combination of various methods: LIS extraction, removal of glycopeptides by WGA-Sepharose, and finally separation of the residual integral proteins by gel filtration on Sepharose 6B in the presence of SDS. These integral proteins, with the possible exception of band 7 and of some minor components, appear to span the membrane, and with the exception of band 7, to be glycoproteins. This has been clearly demonstrated for the major glycopeptide PAS_1, containing about 60 per cent carbohydrate by dry weight, band 3 containing between 2 and 10 per cent carbohydrate, and for the minor components, which could be labelled with tritium after galactose oxidase treatment as shown by Steck and his coworkers.

The next part of the discussion will describe some structural features of the major glycoprotein (glycophorin), as isolated by the LIS-phenol method. The current model of thi s membrane glycoprotein distinguishes three portions - the N-terminal carbohydrate containing segment, a hydrophobic segment of about 20 amino acids and the C-terminal proline-rich, very acidic segment. Only the hydrophobic segment is thought to interact with lipid within the membrane. All the evidence suggests that the carbohydrate containing portion is exposed to the outside. The C-terminal portion on the other hand, extends into the interior of the membrane.

In Figure 3 the gel pattern of PAS stained glycopeptides is demonstrated. As mentioned previously, four bands can be distinguished, the major band being PAS_1, the peptide wit h the lowest electrophoretic mobility. On the same slide I also show a pattern of the same sample which had been heated

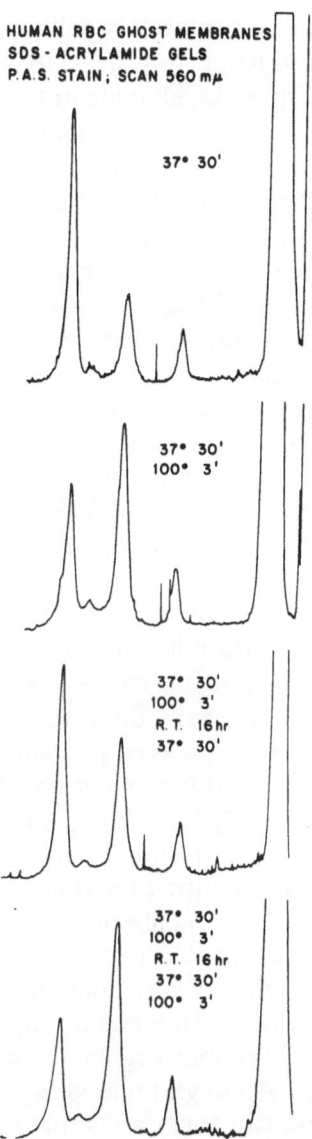

Fig. 3. Effect of heating on the electrophoretic pattern of membrane glycoproteins.

for three minutes at 100°C. This treatment changes the electrophoretic pattern considerably and results in a shift of PAS$_1$ glycopeptides into the position of PAS$_2$. This phenomenon was recently published by Marton and Garvin. (Figure 3b) This change which involves only the glycopeptides PAS$_1$ and PAS$_2$ is dependent upon temperature. A significant difference is seen after incubation around 60 - 65°C, and is also dependent upon the time of incubation at a given temperature. However, this change in

electrophoretic mobility can be completely inhibited by increasing the protein concentration: 100µg of glycoprotein in 50 µl of sample buffer per gel
gives identical results, wehther the glycoprotein-SDS solution had been
incubated at 37^oC or at 100^oC. At concentrations of 1µg/50 µl applied to
the gel, about 80 per cent of the total amount of PAS-staining material migrates in the position of PAS_2. This change in electrophoretic mobility
is reversible, which is shown in Figure 3c. In order to demonstrate this
effect, gel slices containing the glycopeptides PAS_1, PAS_2, or PAS3 were
loaded on another set of gels. Heating of the PAS1 containing slice resulted
upon electrophoresis in two bands, PAS_1 and PAS_2. This result was expected, however, it was not anticipated that upon heating of PAS_2 there
resulted again two bands, one being PAS1 glycopeptide. Without heating,
homogeneous bands for all three glycopeptides could be obtained.

We interpret these results in the following way: incubation of
membranes of the isolated glycoprotein in SDS solutions at 37^oC for 30
minutes is not sufficient to dissociate PAS1 which however occurs after
incubation at higher temperatures. PAS1 thus appears to be a dimer composed of two PAS_2 subunits. The strong dependence on protein concentration suggests that dissociation after heating might be complete, which is
obscured by the method applied, since reassociation seems to occur.

In order to determine where or in which portion of the molecule
this association takes place, we prepared proteolytic peptides from the
isolated glycoprotein. These peptides are derived from different portions
of the molecule, glycopeptides, T_{is} and Chy_{is} (tryptic or chymotryptic
peptide containing the hydrophobic region), and the C-terminal peptide.
When these peptides were mixed with a certain amount of glycoprotein
and applied to gels before or after incubation at 100^oC for three minutes,
the mixture containing the hydrophobic peptides T_{is} and Chy_{is} gave a single band of PAS_2 peptide, which however could only be observed after
heating the sample. (Figure 3d) This result suggests to us that a small
peptide derived from the hydrophobic region can compete and prevent reassociation of the dimer on polyacrylamide gels. If this interpretation is
correct and this glycoprotein can form structures which cannot easily be
dissociated, then we should find some evidence that this molecule does
exist as a dimer in the membrane. However, at this point in time we have
no direct evidence to support this conclusion.

MARCHESI: I think we can now entertain questions or comments about
the study which was presented here or ancillary contributions.

REYNOLDS: Let me point out that the behavior of the glycoprotein in SDS
is readily explained by the binding measurements carried out in our laboratory. Unlike water soluble proteins, the glycoprotein binds very little
SDS below the critical micelle concentration. Above the critical micelle

concentration, extremely large amounts of detergent are bound - up to 5-6 grams per gram of protein. At saturation levels of SDS binding, the glyco-protein is monomeric (polypeptide molecular weight approximately 14,000). Below the critical micelle concentration of SDS, the glycoprotein is aggre-gated. Both the self-association of the glycoprotein and the binding of deter-gent are kinetically slow processes at room temperature. It is important to note that simple mixed micelle formation between the glycoprotein and SDS cannot explain the high level of binding at saturation. One must invoke other factors such as the involvement of the carbohydrate in the interaction with detergent. Furthermore, if the glycoprotein is self-associated within the phospholipid bilayer, one would expect it to be self-associated when insert-ed into a detergent micelle. The major point to be made is that integral membrane proteins are not always well behaved in detergent systems in the sense that they may not bind the detergent in the same manner as water soluble proteins.

UHR: Practically speaking, can you overcome the problem by raising the temperature for a sufficient period of time?

REYNOLDS: Yes

UHR: Then if there are so many problems, why has the SDS polyacrylamide method been relatively effective?

REYNOLDS: Most studies on SDS-protein complexes have been carried out with water soluble proteins. A large number of membrane associated pro-teins are ''extrinsic'' proteins and behave in SDS solutions in a manner sim-ilar to water soluble proteins. Examples of these "extrinsic" proteins are spectrin, and bacterial ATPases. The " intrinsic " membrane proteins, some of which have been shown to have large hydrophobic sequences, are generally not well-behaved in SDS.

KENNEL: What is the size of a micelle of SDS, and is its molecular weight a significant factor in the total protein molecular weight that one finds?

REYNOLDS: The size of the micelle depends upon ionic strength. In water the micelle number is approximately 50. In 0.5 ionic strength it is approx-imately 146. So you have a very wide range. The presence of micelles only affects what is observed on SDS gels if you are binding micelles. As I pointed out previously, water soluble proteins bind only monomers.

MARCHESI: Interestingly enough, althrough there are many problems with the use of the SDS gel technique for determining the molecular weight of glycoproteins, Jackie (Reynolds) has data on the molecular weight of what

she thinks is the monomeric unit of glycophorin which is in the neighbor-
hood of 14,000 daltons for the polypeptide portion. Since we now feel that
there are approximately 130 amino acids per individual polypeptide, and
we can get a value from SDS gels which agrees with this, the agreement
is clearly quite remarkable. This could be fortuitous, but it might also
indicate that the SDS system can really give us reasonable values in spite
of the ambiguity of the method.

CUATRECASAS: What is the possibility that you actually see dimers on
gels but that higher polymers exist in the native state? You have raised
the possibilities of relatively stable complexes in the dimer, but how can
you exclude higher forms? Another question is how many of these would
you need to rationalize the mass of the intramembranous particles were
they to be made up only of glycophorin molecules?

FURTHMAYR: I think you are raising many points at one time. What
I showed in the figure is a minimal model with two peptides associated.
Since we don't know whether this model is true, we cannot say that there
are more than two associated within the membrane. The other question
you raise concerning intramembranous particles is without an answer
since we don't know whether this molecule can form these structures in
the membrane. I think this could be a problem because the number of
residues which associate with the lipid bilayer is about 20 or 21, and if
you assume that there are two of these monomeric structures, you still
only have about 40 amino acids. I am not sure that these 40 amino acids
alone can account for the mass of these particles.

MARCHESI: It is surprising that one never sees larger aggregates of
this material even after repeated heating and cooling. We never see material on top of the gels unless we do things to the material which might
induce cross-linking of the sugar moieties, such as aldehyde formation.
We really do not know why larger aggregates do not form; however, we
have used this as one of the arguments in favor of the idea that the dimer
form may be of some significance. There is a possibility naturally, and
I am sure it is obvious to everyone here, that the dimer form does not
exist in the intact membrane. The molecules could conceivably be all
monomers which are induced to form dimers by some unexplained effect
of SDS.

REYNOLDS: The inside of an SDS micelle is the same as the inside of a
phospholipid bilayer that is fluid. In relation to Cuatrecasas' comments,
there is a more important point here that none of us have looked at. The
carbohydrate end of the glycoprotein and the C-terminal end are negatively
charged. In the membrane these charges may be neutralized by Ca^{++} or

Mg^{++} binding and the macromolecules may self-associate. When we try
to put the glycoprotein into SDS or DOC micelles, we see only monomer
perhaps because of electrostatic repulsion between the highly charged
terminal regions.

MARCHESI: Anyway, we feel it is not entirely safe to conclude that the
internal part of the membrane is like SDS, although it may be. One experiment we did is a variation of what Heinz (Furthmayr) talked about, and
the rationale was that if we dissolved membranes in SDS containing high
concentrations of the particular peptide which was so effective in preventing dimers, then you might extract all the monomer that existed in the
membrane. The result is that membranes dissolved in SDS containing a
high concentration of the competing peptide produce exactly the same SDS
gel pattern as a normal membrane. If they are heated in the presence of
the competing peptides, then one obtains all monomers or pseudomonomers.
Although this experiment is not conclusive, it would suggest that the molecule may even exist in the membrane as a dimer.

EDELMAN: Jackie (Reynolds) you raised a point about the charge, but
isn't there also an excluded volume effect for the carbohydrate chains in
a glycoprotein? Has anyone measured the excluded volume and estimately
how this would influence the interactions of these molecules?

REYNOLDS: Work of this sort has been done on branched polymers by
Florey. One might want to measure this effect in the glycoprotein, but to
my knowledge it has not been done.

JEANLOZ: I would say that I am not really informed about the problem
of the excluded volume, but I have a vague feeling that the work has been
done in the past, probably on pure carbohydrates. If you were to go back
to the thirties, work was done on pure carbohydrates, which provide a
good model for the branched polysaccharides of glycoproteins.

MARCHESI: There is really no data available right now?

JEANLOZ: Maybe, but you don't know if it exists.

MARCHESI: We worried about the association between dimers being by
virtue of the sugar, not the hydrophobic region. We thought that the hydrophobic region would be the least likely to maintain associations in SDS,
and thought that something funny was going on with the sugars. Competition experiments with sugar antagonists suggested that they had no effect
on the dimer.

BRANTON: I want to comment on some other remarks that have been made here regarding charge repulsion. If you have two charged molecules and wonder whether they might be a dimer inside the membrane, I think it is important to define your salt concentration. Obviously, at physiological salt concentration, charge repulsion is not going to be terribly important unless you are distinguishing between very small distances.

JEANLOZ: I think it may be time to discuss the question of aggregation between carbohydrate molecules, because it is a subject that has been left out for at least thirty years. There was a great controversy between Staudinger, who supported the concept of non-associated macromolecules and Kurt Meyer who supported the concept of micelle formation. These micelles could be formed by macromolecules especially polysaccharides. At the time I was working in the laboratory of Meyer and we were attempting to dissociate micelles with chloral hydrate, a reagent which will be discussed here later (Session IV). I think some work should be done again on the aggregation of carbohydrate chains versus the aggregation of protein chains. To my knowledge still little is known about the conditions necessary to aggregate two carbohydrate chains.

MARCHESI: We have been trying to find out something about this, and you are right the literature is uninformative. How do you propose to study it? What can you use?

JEANLOZ: I would propose simple models like amylose, amylopectin, and glycogen.

MARCHESI: Are they simple models? Is their structure known? We face the danger that these appear simple, but they may be complex.

JEANLOZ: Because we have only one type of residue, the D-glucose residue, only one or two types of branching [α-(1— 4) and α - (1 —— 6)], and we know the molecular weight exactly, I feel that enough is known to study the physical parameters of aggregation.

REYNOLDS: Let me say in relation to this that in 1963 Reynolds and Halsey published a study of aggregation in a branched polysaccharide from locust bean. It is an uncharged polysaccharide which can only be disaggregated in aqueous media by acetylation of the hydroxyl groups.

HAKAMORI: Does the dimer exist as a helix in the membrane?

MARCHESI: I was not pushing for the helix. We have no idea what it really is. Wouldn't the helix be unlikely if the molecules were dimers?

SONENBERG: It would be better if we stabilized it.

MARCHESI: Are you saying this is an argument against the dimer being a helix in the membrane?

SONENBERG: I think so.

MARCHESI: Dr. Tomita is here and can discuss some of our structural studies which should present some other interesting points about the location of the charged groups.

TOMITA: Our amino acid sequence analyses, radio-iodination studies, tryptic peptide and cyanogen bromide peptide analyses all indicate that the glycophorin molecule consists essentially of three segments. The carbohydrate containing N-terminal portion of the molecule is the most hydrophilic and is exposed at the plasma surface. A central portion of the molecule consists of a sequence of very hydrophobic amino acids which is contained within the lipid bilayer. Finally a somewhat more hydrophilic C-terminal sequence extends into the cytoplasm. Our preliminary data indicate that the C-terminal portion of this glycoprotein is phosphorylated, possibly on serine and threonine residues.

MARCHESI: The point about tyrosine is very important because it does seem to be iodinatable, and this was somewhat unfortunate because we had already used the tyrosine plus three other amino acids on the inside of the membrane spanning portion. So this is sort of making it more restrictive as to how much polypeptide chain might really be within the bilayer. We really don't know this, but now we have gone down to less than 20 residues. I just wanted to emphasize again that the carboxy-terminal end is made up of acidic amino acids, in addition to some phosphothreonine and phosphoserine. Since the phosphorylation is due to endogenous activity, this is the most compelling evidence that part of the molecule has to be inside the cell where the kinase is located. When we get onto the question of what is hooked up to this molecule keep in mind a rather curious distribution of charged residues. Any comments?

PINCUS: With respect to what you just said about phosphorylation of glycophorin, do you have any evidence whether the degree of phosphorylation of the protein affects the monomer-dimer interconversion?

MARCHESI: No, we don't.   What we do know is that every glycophorin
is phosphorylated - everybody working with this molecule has been working with a phosphorylated form.   This is the way it comes out of the cell
and we have not learned how to control the phosphorylation.

PINCUS: What is the success of phosphorylation in the glycophorin isolated?

MARCHESI: We don't know that.   We know that if the cell is fed P^{32}
inorganic phosphate, P^{32} is found on the residues derived from the last
twenty amino acids of this molecule exclusively.   Comments on the sequence or ideas as to what the conformation might be?

REYNOLDS: While you are looking at sequences, can I show you another
one which I think might be interesting?   This is the sequence of the coat
protein from FD virus.   The important region is a central portion of 18
amino acid residues which contains no charged groups.   Drs. R. Webster,
S. Makins, and C. Tanford have studied this protein in DOC and SDS. The
protein forms mixed micelles with both detergents, but in dimeric form.
That is, a dimer of 10,000 molecular weight binds one detergent micelle.
This is in direct contrast to the glycoprotein which is monomeric in these
systems.

FURTHMAYR: SDS-polyacrylamide gel electrophoresis of the hydrophobic peptide T_{is} gave a molecular weight of 6000 - 8000 daltons, indicating that this peptide actually is a dimer which cannot be dissociated.

MARCHESI: The data which Dr. Reynolds just presented seems consistent with the behavior of individual hydrophobic peptides also dissolved
in SDS as shown by Furthmayr.

SONENBERG: I should like to discuss the hydrophobic portion of your
peptide, and comment about its possible conformation.   There are about
20 amino acids in that region.   It is true that the peptide has a lot of hydrophobic residues that Tomita showed.   However, an analysis in terms of
helix breakers and helix formers indicates that there are other factors
which may determine whether you end up with a helix.   For example, the
hydrophobic portion of the peptide has a lot of glycine which is a helix
breaker, and threonine which is indifferent as a helix former.   From an
analysis of the 15 proteins whose structure has been determined by X-Ray
crystallography, the rules suggested by Chow and Forsman would indicate
that you have little helix in the hydrophobic region.

EDELMAN: Extremely hazardous. I wonder if anyone knows of any studies of helical structure or β structure in hydrophobic solvents, particularly in systems like these. It seems to me that you can't make reasonable guesses just by looking at the sequence of these molecules. We gave the sequence of Concanavalin A to some people interested in the prediction of conformation and they guessed 30 per cent helix. The molecule contains much β structure but no helix. So except for the fun involved in it, it is not very productive.

YAHARA: I would like to make a brief comment relevant to Dr. Edelman's comment. I think that the conformation of a given polypeptide is determined by its amino acid sequence in a given medium. If polypeptide exists in a usual aqueous medium, the parts of the polypeptide chain that consist mainly of hydrophobic or non-polar amino acids could be folded inside the molecule. But the case is different when this polypeptide is dissolved in a non-polar medium, for example some kinds of organic solvents. In other words, even when the sequence of the hydrophobic region of glycophorin was determined, we cannot predict its conformation according to knowledge based on local conformations of water soluble proteins which have been analyzed for amino acid sequence.

MARCHESI: Would you say that the inside of a protein or glyco protein is hydrophobic?

YAHARA: What I would say is if you put a known sequenced protein in aqueous medium, its hydrophobic region could be folded inside. But at the same time the outside amino acids affect the inside structure. On the other hand, when you put the hydrophobic region alone in a non-polar medium, a way by which we estimate its conformation may not be obtained from the knowledge of the relations between the amino acid sequences and conformations of water soluble proteins. Since the hydrophobic region of glycophorin is located in lipid, this is obviously one of the latter cases.

CUATRECASAS: I just wanted to follow up on earlier comments regarding aggregation in liposomes. Given the propensity for aggregation of membrane proteins even in the presence of detergent, a very difficult, but very important, question is the forces or mechanisms keeping all the proteins from coming together on the membrane into a big glob containing all the proteins? You have not mentioned phospholipids but in your mind what might be the role of phospholipids particularly those highly ordered in a shell around the membrane, which may act to strongly restrict adhesive intermolecular interactions.

MARCHESI: I guess what Pedro is talking about is that there exists around the molecule certain amounts of tightly bound lipid which must

serve as punctuation marks. One idea, originally pointed out by Griffith
and Jost, and mentioned by Cuatrecasas, is that if these hydrophobic segments are so keen on associating together, why isn't the cell membrane
just a glob of proteins stuck in some corner somewhere? It has been suggested that this region of the molecule is coated with some very special
lipid which has a much higher affinity for that segment of polypeptide chain
than it does for other lipids. That's what keeps the proteins apart. You
must think of it in terms of being like a myelin sheath around each individual polypeptide. An interesting idea!

McCONNELL: That is a fascinating subject. There is some lipid around
glycophorin for instance, and its physical and chemical state depends in a
very sensitive way on the lipid composition of the medium. It is highly
temperature dependent. It depends on whether or not there exists solid
lipids in equilibrium with fluid lipids at a specific temperature. Also the
lipid structure around glycophorin is probably asymmetric, and very likely there is more solid lipid on one "side" than on the other "side".

MARCHESI: You feel as a generalization the boundary lipid idea is a real
one and has to be considered?

McCONNELL: It can be detected in very sophisticated experiments. It
is hard for me to imagine it should not affect the polymerization of glyco-
phorins for instance.

MARCHESI: Turning the question around, do you think it is necessary to
have bound lipids to keep these molecules apart?

McCONNELL: We don't know yet if there is a correlation between polymerization and the degree of bound lipid, but I would be surprised if the
degree of bound lipid, which is temperature and composition dependent,
did not affect the polymerization of these molecules.

MARCHESI: Any more comments? What we have tried to do is to establish the principle that there are buried within cell membranes (certainly
the red cell) protein molecules which have a kind of transmembrane conformation. There are two interesting questions which follow from this
model: what happens when a ligand binds the external surface, and what
does it do for/to the cell? There is a popular fantasy that stimulation on
the surface causes some kind of signal to be transmitted into the cell. A
lot of work in the past few years has focused on the whole question of what
goes on inside the cell, what are the connections between these proteins
on the surface and the other proteins in the membrane.

BRANTON: I will be speaking about largely structural approaches to the question of protein-protein interaction. Other ways in which one could examine the interactions between proteins include for example cross-linking studies, a problem, which I am sure Dr. Steck will address. We have spoken also of aggregation of proteins, and certainly the ways in which proteins are bound to each other within the membrane vis- a - vis the mobility of the individual proteins. If certain proteins that are, for example, fibrous in nature can form a network, as appears to be the case in the red blood cell studies which I will be speaking to you about, it is possible that such a meshwork can limit the mobility of other proteins in the membrane.

The studies I will be speaking to you about refer to red blood cell ghosts. These are studies that have been done in collaboration with Arnljot Elgsaeter and David Shotton. We have been examining red blood cell ghosts primarily by freeze-fracture techniques. I assume that most of you are familiar with these techniques.

The particles which one sees in red blood cell membranes are almost certainly protein, as judged from the fast that one can recon-stitute a wide variety of different hydrophobic proteins into lipid bilayers and in effect reconstitute this kind of image of a particulate surface. They probably represent the glycoprotein and the Band 3 proteins which you heard about earlier this morning, although a clear demonstration of the relationship between a given protein and the particle, I think still has to be made.

The remarkable thing about the distribution of the particles is that in fresh cells or in ghosts, they generally appear in a fairly random distribution, but under certain conditions one can cause them to aggregate. Figure 4 shows this kind of aggregation. This is a partial aggregation in-duced by changing pH value; it is reversible. One can demonstrate a more aggregated state. Such aggregation cannot be simply induced by changing the pH value. It can be induced by the application of either calcium or positively charged proteins. Again, this type of aggregation is revers-ible. The key observation is that this kind of aggregation cannot be induced, or at least we have not been able to induce it, in a fresh cell or a freshly prepared ghost. The only way in which we have been able to induce it is with low p H values or with positively charged proteins.

The proteins which are actually removed from the red blood cell ghosts as a result of an effective pretreatment. As a result of extrac-ting spectrin and Band 5 we can induce the kind of particle aggregation I referred to previously. It is not some co-incidental effect of the incuba-tion, and there are various kinds of experiments which show this. For example, one can cross-link spectrin and Band 5 to the red cell so that it fails to come off, for example by cross-linkage by glutaraldehyde treatment. If we do this and pre-treat the cells in a way which normally causes the

removal of these proteins, then we do not obtain intense aggregation.
In other words, if we cannot remove these proteins during an incubation
which would otherwise remove spectrin and Band 5, then we fail to get
subsequent aggregation. We have also been able to show that there is a
quantitative relationship between the amount of spectrin and Band 5 that
are removed, and out ability to aggregate the particle. What is implied
again is that it is a phenomenon independent of the particular conditions
used to remove spectrin and Band 5, but dependent only on the amounts
of the peptides removed. The molecular mechanism is not clearly understood, but we believe that there is a meshwork of spectrin and Band 5
which in the fresh ghost limits the mobility of these particles either by
specifically binding to the particles, or to the protein of the particles,
or by physically entrapping the particles.

An obvious question is whether there is sufficient spectrin to
do this. One can perform some calculations and develop a scale model
which we have done and presented elsewhere (as a photograph of model)

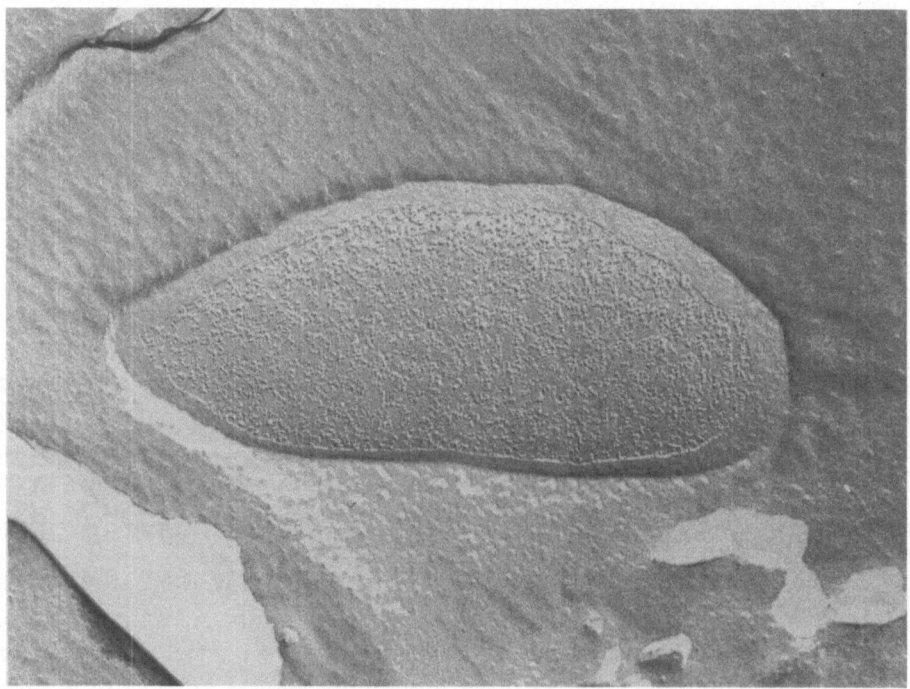

Figure 4: Freeze etch view of a human erythrocyte ghost showing normal,
dispersed distribution of intramembrane particles.

using balls and pieces of string. If you assume that the spectrin is plastered up against the membrane, and in fact we do have electron microscopic evidence that this is the case, there could be extensive cross-
linking between spectrin and membrane particles. It is rather interesting that if one tries to induce aggregation of particles, as we have seen
it, in ghosts where we have not removed the spectrin, instead of getting
particle aggregation one observes a very different phenomenon. One
sees blebbing of the red cell ghosts and only partial aggregation. These
blebbings are formed as one aggregates the spectrin by adding calcium
or positively charged protein to the inside of the membrane. These bleb-
ings are formed completely free of any of the particles and, in fact, if
one attaches by diazotization phenyl-lactose, we can see that the blebs
that come out of the cell are completely free of any of these lactose determinants. (Figure 5)

Figure 5: Protamine induced blebbing of the erythrocyte ghost. None of
the ferritin anti-lac (F) binds to the bleb which appears to contain only lipid.

This is remarkable, because there is obviously a mass flow of lipid material past the particles in order to form this bleb, and yet the particles are held in place. We explain these observations in the following way. (Figure 6) Our results showing that spectrin molecules limit the mobility of the erythrocyte ghost intramembrane particles may have important implications for the structural arrangement of other cell membranes. The association between spectrin and the particles, which are known to extend to the outer surface of the membrane, suggests a means whereby proteins within a cell could affect the distribution of cell surface components. Although abundant quantities of spectrin have not been detected on other cell membranes, a similar association between intrinsic membrane components and an underlying network of fibrous or filament-forming protein may be a common mechanism whereby the mobility of surface macromolecules is controlled by other cellular constituents. The association between spectrin and the particles may have important implications for the structural arrangements of other cell membranes. The association with the particles, which are known to extend to the outer surface of the membrane, suggests a means whereby proteins within a cell could affect the distribution of cell surface components. Although abundant quantities of spectrin have not been detected on other cell membranes, a similar association between intrinsic membrane components and an underlying network of fibrous or filament-forming protein may be a common mechanism whereby the mobility of surface macromolecules is controlled by other cellular constituents.

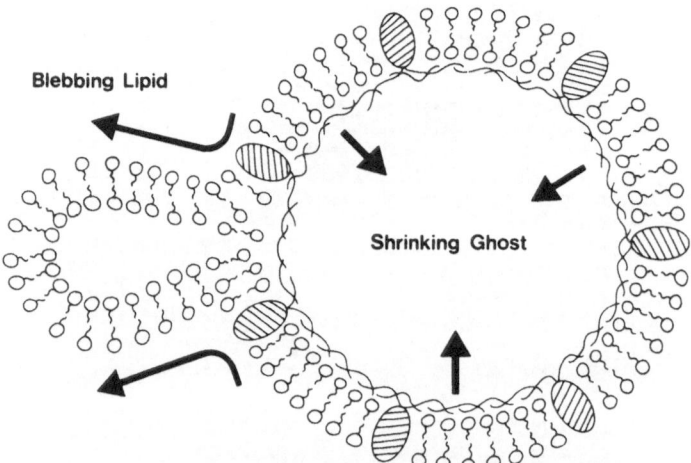

Figure 6: As spectrin meshwork contracts, surface area of ghost falls below minimum for lipid bilayer, and lipid squeezed out as a bleb. Because bleb contains no protein or particles, mobility of membrane particles must be constrained by the spectrin meshwork.

MARCHESI: This is really a very provocative model of Branton's.

EDELMAN: I wonder if you could draw a profile view with the specific question in mind as follows: how big is the particle? Do you propose the particles are in some depth in respect to the outer surface of the membrane, and are you proposing that in fact the spectrin goes into the membrane or the particle interrupts the inner lamella of the membrane? Can you draw a side view in terms of what you know about the physical dimensions, just to fill out the model, because I don't see the whole picture?

BRANTON: I can draw just about anything because we know very little.

EDELMAN: That gives you a lot of leeway.

BRANTON: It is worth pointing out clearly that we cannot on the basis of freeze fracture say a great deal about the precise dimension of the proteins involved in the particles. There is a real problem when it comes to establishing the extent to which the particles might extend beyond the lipid bilayer.

EDELMAN: I will tell you why I asked the question. It seems to me that one of the key facts we lack has to do with the rate of aggregation of the particles. If, for example, the particles were entrapped completely in the lipid bilayer, you would expect one rate of aggregation involving one set of forces. Whereas, if you had the kind of interaction you proposed, this rate would be perturbed in a predictable manner. I wonder if anyone has ever designed an experiment that would give information on this rate of migration or clustering of the particles.

BRANTON: I don't know of any specific experiments that answer that question. A. Elgsaeter in my laboratory has designed an experiment that could perhaps attack this problem. The general notion is to label the components of the particles in which we are interested with fluorescent antibody at random throughout the cell, and then bleach out the fluorescence from a particular portion of the cell. The rate at which the fluorescence migrates back into the bleached region would then be a direct measure of mobility.

KARNOVSKY: Two questions. In the pictures where you had quite a massive aggregation, there was a certain population of particles that does not aggregate. Is this just a variation?

BRANTON: That is an interesting point. I have gone into the mechanism of aggregation. In fact, we suspect that the aggregation itself may be caused by an aggregation of the residual spectrin that remains on the surface membrane. When you remove enough of it so that there are particles sitting there with no spectrin attached, this might explain why they remain disaggregated.

KARNOVSKY: Another one about aggregating spectrin: when you add high calcium you get attenuation from the lipids, or are you aggregating the spectrin in that situation.

BRANTON: What we have done is this: we have isolated spectrin and measured in isolation its precipitation under all varieties of conditions which we then used in the cell. We assume, and this is an important assumption, that it is aggregating in the same way when attached to the membrane. So the answer is, when we thow in calcium, I assume we are aggregating the spectrin. When we throw in positively charged proteins I assume we are aggregating the spectrin. There is some evidence from Dr. Richard's laboratory showing extensive cross-linking of all of the proteins to the spectrin. This is very much the same kind of thing we observed, in effect, when we aggregate the spectrin and release the bleb; all of the proteins are excluded from the bleb, suggesting the spectrin is able to entrap or prevent the movement of all the other proteins. I think we will have to wait and see how the dust settles on all the various cross-linking experiments before drawing any definitive conclusions. Perhaps Steck would like to speak to this.

STECK: I would like to summarize our views on the relationship of the red cell membrane fibrillar proteins to the membrane-penetrating glyco-proteins. As depicted in Figure 7, Band 1 and 2, called spectrin, are polypeptides of more than 200,000 daltons molecular weight which resemble myosin in amino acid composition. They are selectively eluted from the membrane at low ionic strength along with band 5, a 45,000 dalton species which resembles actin in its composition. Guidotti has invoked these similarities to suggest that Bands 1, 2, and 5 may be a red cell membrane "actomyosin". Sheetz, Painter, and Singer have recently strengthened this analogy by demonstrating that purified Band 5 can be polymerized into fibrils which which stimulate myosin ATPase and which can be decorated by heavy meromyosin to form an arrowhead pattern. These three fibrillar polypeptides are clearly confined to the cytoplasmic surface of ghost membranes. The two glycoproteins which we believe span the ghost membrane are Band 3 and band PAS_1. When ghosts are extracted under defined conditions by Triton X-100 solutions, these two species along with about half of the membrane proteins and most of the phospholipid are selectively solubilized. The insoluble pellet contains predominantly Bands 1, 2 and 5, but also other cytoplasmic-surface polypeptides.

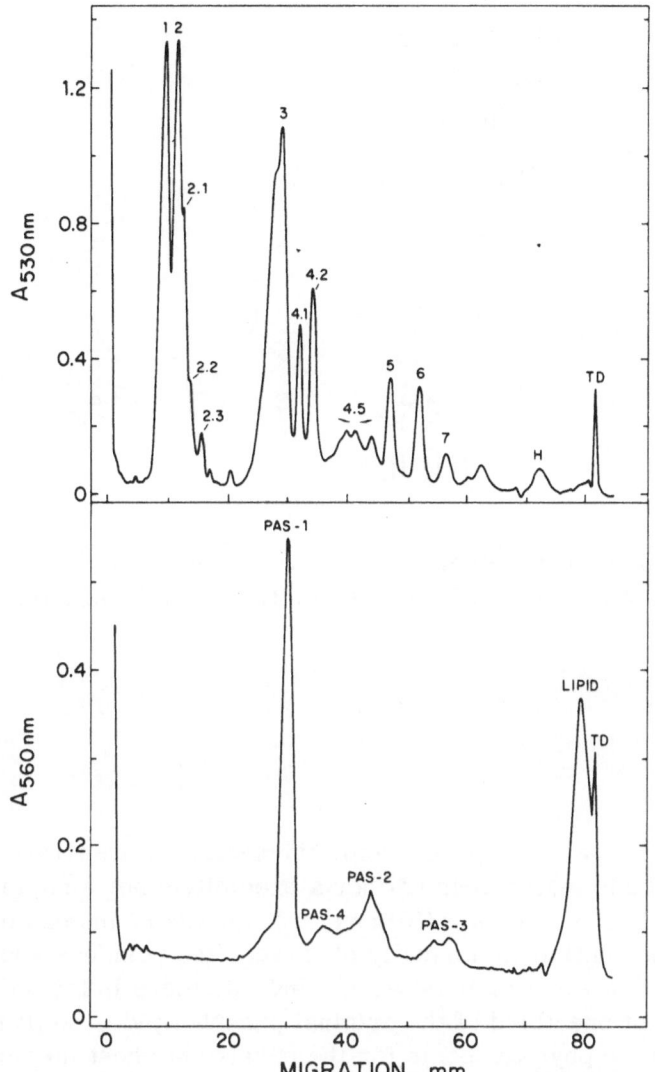

Figure 7: The electrophoretic profile of red cell membrane polypeptides on a polyacrylamide gel in sodium dodecyl sulfate. Upper panel: Coomassie blue stain. Lower panel: periodic acid Schiff (PAS) stain. From T.L. Steck, J. Cell Biol., in press (July, 1974).

Figure 8 depicts a sketch which sums up our current understanding of the topography of these major polypeptides on the red cell membrane. Along with Donald Fischman and John Yu, I have examined by electron microscopy the pelleted material remaining after Triton X-100

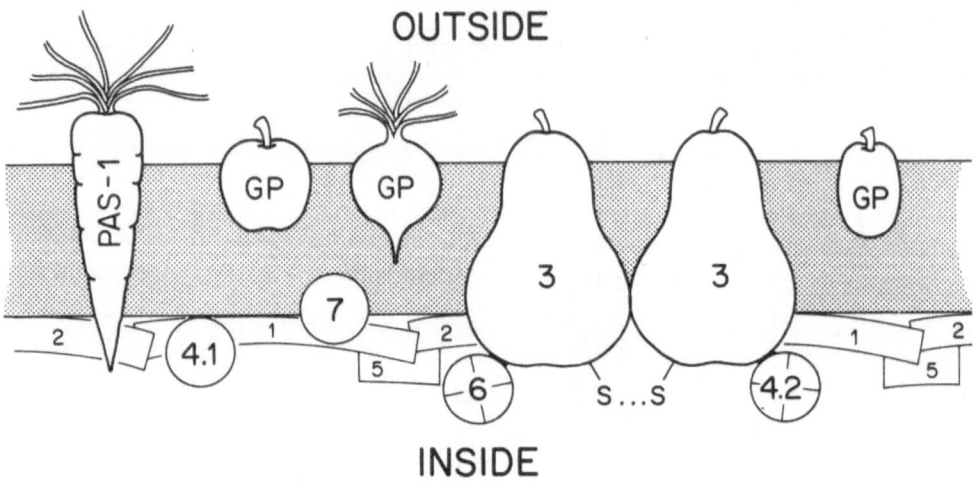

Figure 8: A possible arrangement for the major polypeptides of the human erythrocyte membrane. From T. L. Steck, J. Cell Biol. July 1974, in press

extraction. In both negatively stained and thin-sectioned preparations, we find that the insoluble material is neither a precipitate nor an aggregate, but discrete residues of ghosts. (Figure 9) However, instead of the smooth membrane continuum normally observed, the residues appear to be held together by a fibrillar meshwork. Indeed, there is left in these residues only about one-third of the original phospholipids, so that a lipid bilayer cannot be the physical basis for the persistent ghost membrane structures observed. Instead it is the matrix of fibrils in the form of a self-associated continuum, which holds the residue together. The lipid remnants remaining appear to form small segments of bilayer which remain associated with the fibrils. So we imagine that spectrin forms a thin continuous sub-membrane reticulum which confers mechanical stability to the red blood cell. Branton (and also Nicolson) have raised the hypothesis that the spectrin fibrils could be binding to the cytoplasmic aspects of the two membrane-penetrating glycoproteins, Bands 3 and PAS$_1$, and thereby restricting their freedom of motion in the plane of the membrane. I would suggest that it seems more likely at the present that the spectrin does not actually bind these penetrating proteins but merely constrains their motion by entrapping them in the interstices of a dense fibrillar reticulum just discussed. The reason for preferring the latter hypothesis is merely that we have not observed a physical association of Bands 1, 2, and/or 5 with either Band 3 or PAS_1 in non-denatured, Triton X-100 solutions. Under these conditions, Band 3 still retains its dimeric structure and appears to bind Bands 4, 2, and 6, as it presumably does in the ghost. If spectrin could bind Band 3 or PAS_1 we would expect to see it.

In a similar vein, we do not observe any association between Bands 3 and PAS_1 in Triton X-100 solutions, although it has been suggested that these two species could be associated to form the intramembrane particle revealed by freeze-fracture electron microscopy. Our cross-linking studies have likewise failed to reveal Band 3 - PAS_1 complexes. I believe that Band 3 could, of itself, account for most of the freeze-fracture particles, since as dimers of approximately 180,000 daltons, it could provide the observed approximately 500,000 membrane-spanning particles of about 78 Å diameter.

MARCHESI: We should not stop here, however it is time for the morning break. We shall have a recess now, but will take up this issue when we come back.

COFFEE   BREAK

MARCHESI: This is important, because I am sure that immunologists and cell biologists who are looking at movement and mobility of components will have trouble if they must worry about connections inside the cell. Let me say that as of this moment, there is no evidence that there are proteins outside the cell that are organized this way. The generalization that has emerged from studies of freeze-etching and electron microscopy is that there are structures inside the lipids. Branton is suggesting that this globular structure is a kind of interface that goes from the outside surface to the inside surface of the membrane. He knows that there are particles, maybe a quarter of the total, that are much closer to the outside surface of the cell than they are to the inside. In fact, if one postulates the connection between these particles and the inside, you have to almost invoke some new ideas, and especially, if you get back to what Edelman wants to know, which is how close are the spectrin molecules to the lipid elements in the membrane. We do not know; and this is of importance. There may be two populations of particles and most of us are ignoring these populations, because it is technically somewhat difficult to study them quantitatively. But this is an important question when one considers other cells. Almost invariably people ignore this population of particles.

Let's leave that for a second. Maybe Branton wants to come back and answer that point. The generalization seems to be that the two major integral proteins are in some way related to thi s globular unit. Steck

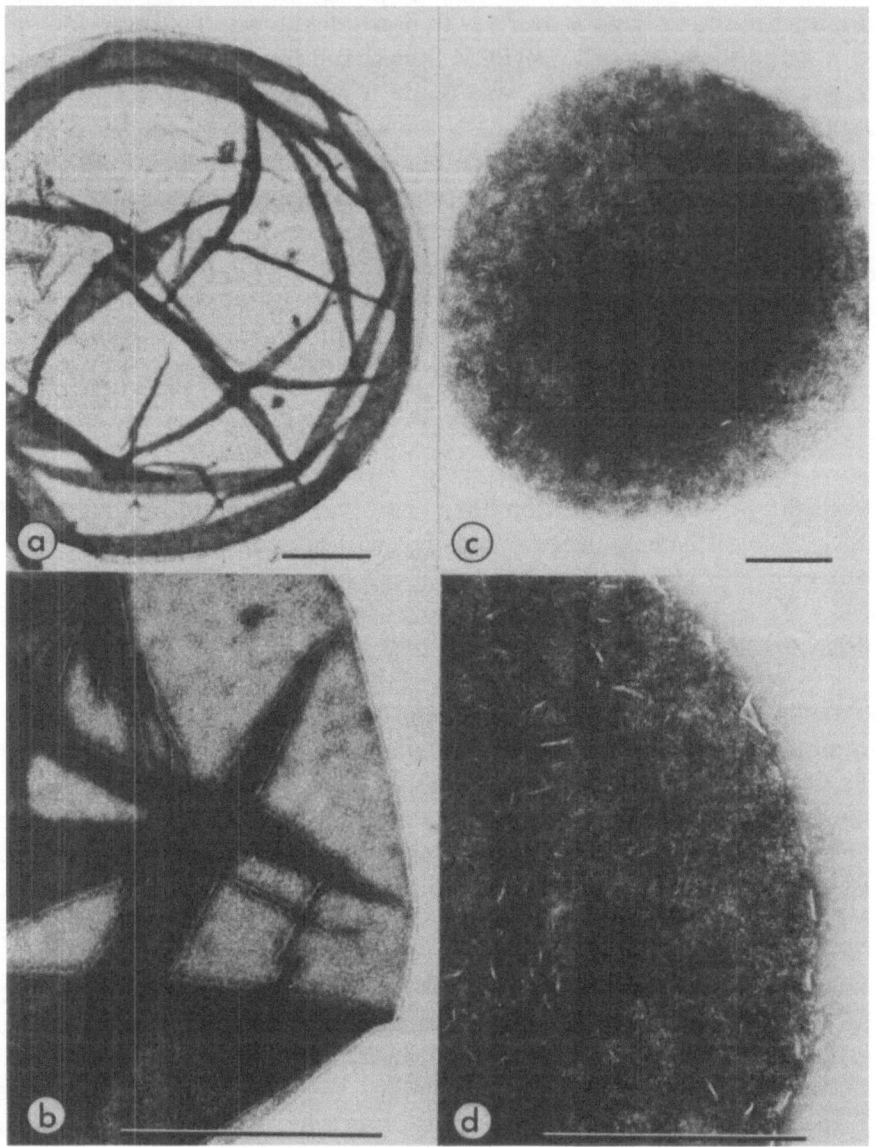

Figure 9: The fibrillar residue of ghosts following extraction by Triton
X-100. Electron micrographs of negatively stained control (a and b),
and 0. 5 per cent Triton X-100 extracted (c and d) ghosts. Calibration
bar is one micron.
From: J. Yu, D. A. Fischman, and T. L. Steck, J. Supramolecular Structure 1: 233, 1973.

feels that since Band 3 is 90,000 daltons and therefore may be something like 900 amino acids, it would certainly have enough mass to form this unit. I think he has to explain why he feels that it is appropriate to put Band 3 in this form, rather than just simply make it a globular unit which is largely inside the cell. He probably has some data to support that, but I want to point out that there is no simple way that we can pack these molecules in the membrane. In fact, we have not eliminated the possibility that a certain amount of the glycoprotein molecules are really in the outer half of the membrane exclusively, and never reach the inside. I caution you to recall that all the labelling data is largely qualitative. No one has yet been able to say how much of these molecules are labelled under the different modes. If only ten per cent of the mass of these molecules really are distributed inside the cell membrane, the way the study has been done up to now, one would never know this.

Before we proceed with other discussions, it might be appropriate for Steck and Branton to come back to these points. First, comments from Branton concerning the possible heterogeneity of the particles and what they mean to the model.

BRANTON: I don't know that there is a great deal of heterogeneity established in the case of the red blood cell membrane. In other words, I do not know whether the particles are heterogeneous or homogeneous. There is nothing in their morphology that would suggest heterogeneity at this stage of resolution and nothing in the way of labelling studies to suggest heterogeneity.

You do point out, I guess, in your diagram that some of the particles can go to the outside of the membrane and some inside. The question obviously arises, is there a fraction of particles that are attached to the outside as against those attached to the inside. I think there really are no studies which would answer that question directly in the case of red blood cells. There is a rather interesting approach. B. Satir studied the question in Tetrahymena. In this microorganism there are rosettes of particles. Any particle in the rosettes appears to have the same probability of landing on one or the other side of the membrane as its neighbors in the rosette. Hence, all the particles in the rosette are equal; there is no evidence of heterogeneity.

CUATRECASAS: Have you by any chance studied particles of freeze-fractured materials using anti-spectrin antibodies analogous to the work of Nicolson?

BRANTON: Nicolson has used highly purified anti-spectrin and has found that when anti-spectrin is applied to the inside of the red blood cell ghost,

it can induce aggregation of negatively charged sites at the outer surface
of the membrane. These negatively charged sites are associated with
sialic acid, and if we believe the sialic acid is in turn associated with
glycoprotein which is associated with the particle, then you might assume
that what has been shown in effect is an aggregation of the particle. We
have done similar experiments and do see some aggregation of the particles upon application of anti-spectrin antibody, but it is not nearly as
marked as I would have been led to expect on the basis of the pictures
that Nicolson has shown in which there are very obvious patching of negatively charged sites. So whether or not he has moved all the particles,
I really do not know. There is some particle movement when you apply
anti-spectrin antibody, but not as great as I would have expected. There
is another problem associated with the experiments. Nicolson's sequence
of manipulations probably leads to the removal of some of the spectrin.

MARCHESI: While you are here, could you comment on the evidence that
spectrin really exists up against the membrane in intact cells. One of the
ideas we have is that spectrin might very well be dipping into the cell, possibly serving as a scaffolding for enzyme complexes or for some other
reasons.

BRANTON: I think you are asking: What is the evidence that spectrin is
plastered up against the membrane rather than dangling inside the cell?
In the native red blood cell membrane, I don't know of any evidence right
now. I think the best approach to that would be to apply ferritin-labelled
antibody to sections, using Singer's frozen sectioning techniques. When
we examine freshly prepared ghosts, we do not see any spectrin dangling
from the membrane. We can visualize spectrin molecules very easily
when it is isolated in solution, so there is no difficulty in our ability to
see spectrin. We don't see it on the native membrane. On the other hand,
we can see fibers peeling off the membrane seconds after we place the
membrane in low ionic strength medium, where the spectrin is measur-
ably released from the membrane. This evidence suggests that spectrin
really is plastered up against the membrane. It accounts for the fact that
we can't see it in the fresh ghost, because it is so close to the membrane.

MARCHESI: I hate to dominate the discussion, but one more question.
You said before that there is no evidence for substantial amounts of spectrin in other cells. How do you arrive at that conclusion?

BRANTON: I don't know who said it. I did't say it was not there. I
said there is no evidence for it being there, and could you supply me with
the evidence.

MARCHESI: I thought you implied that there is good evidence - so the lack of evidence means that the question is open.

STECK: With regard to the last point, Guidotti has argued that "spectrin-like " proteins may exist in many types of plasma membranes although systematic data have not been published.   Neville's "eigen-protein" may also fall into this category.

   With regard to the red cell membrane glycoproteins, I would not preclude at this time the possibility that certain integral glycoproteins are present at the outer surface but do not penetrate through to the cytoplasmic surface. This question deserves systematic study. In the case of Band 3, I think the evidence is strong that essentially every copy of the glycoprotein extends asymmetrically across the plane of the mem - brane. Thus every band 3 polypeptide is digested at both the outer and the inner (that is, cytoplasmic) surface of the membrane, but the digestion pattern is different at each side. We also know that there are binding sites for glyceraldehyde-3-phosphate dehydrogenase on the cytoplasmic aspect of Band 3, and that the stoichiometry is such that each Band 3 polypeptide can bind one glyceraldehyde-3- phosphate dehydrogenase molecule. Every Band 3 polypeptide can be cross-linked into a disulfide-linked dimer by gentle oxidation - only at the cytoplasmic surface. These data together with a variety of labelling studies (which I agree may be limited in not tagging every copy of Band 3) all indicate that every copy of Band 3 faces both membrane surfaces, but asymmetrically with the sugar side out.

   Finally, with respect to the question of which hemi-membrane will bear the intra-membrane particles, I want to suggest an hypothesis. Perhaps if the center of mass of such penetrating proteins lies toward the cytoplasmic side of the bilayer, it will remain in the monolayer following freeze-fracture.   For example, Band 3 might be centered toward the cytoplasmic stratum, while PAS_1 (glycophorin) might be held in the external side of the bilayer (in keeping with Marchesi's sequencing data).

MARCHESI: How much mass is in this lipid? How do you go about finding that?

STECK: With regard to the disposition of Band 3 in the membrane, most of the polypeptide can be readily digested in the ghost membrane, suggesting that only a small fraction is hidden in the hydrophobic core.   Since pronase attack on intact cells generates a 38,000 - 40,000 dalton fragment of Band 3, one might deduce that at least much of the 90,000 dalton polypeptide is exposed at the external surface.   However, this topic requires careful investigation.

POLLARD: I have one response to your proposal that unattached glycoproteins may exist in the outer aspect of the phospholipid bilayer. Branton's
experiments suggest that red cell membranes form pure phospholipid blebs
when calcium is added. Possibly glycoproteins floating free in the outside
of the bilayer ought to be concentrated in those blebs. If you could concentrate them enough, you might be able to find the glycoproteins. But in fact,
there isn't any protein in them, according to Branton. Has he run gels on
the isolated blebs?

MARCHESI: He doesn't know that. He says there is none in the particle?

BRANTON: If you analyze the lipid vesicles, there is essentially no protein. There is always some contamination, but I would say that it is very,
very improbable that there is a significant amount of glycoprotein concentrated in those lipid vesicles.

POLLARD: When you run gels of those isolated blebs, do you see a small
amount of those isolated proteins that would be found in the gel of bleb cells
anyway?

BRANTON: If we do gels? We see essentially nothing in the gels, except
a smear of the lipids at the bottom, a very big smear, because we put a
lot of lipid in.

MARCHESI: That suggests the possibility is less likely. Johnathan (UHR)
do you have something to suggest about other possibilities?

UHR: The thrust of my comments is that there appear to be two classes
of membrane macromolecules on the surface of lymphocytes: one class
which appears to be deeply embedded in the membrane, and the other
which is more superficially embedded. Further, there appear to be restrictions in the movement of both classes, particularly the one that appears
to go deeply into the membrane. The experimental evidence that leads to
these ideas was gathered by Dr. Ellen Vitetta. I stress that these are ideas
and not conclusions. The molecules which I am discussing as representative of these two classes are immunoglobulin and H-2 alloantigens, the
major histocompatibility antigens of mice. Both of these molecules are
presumed to be recognition units on the surface of lymphocytes. If one
enzymatically radio-iodinates intact mouse splenocytes, both molecules
on the surface of cells are labelled. Incubation of such splenocytes for
approximately six hours in vitro, during which time there is no evidence
of cell death, releases about 50 per cent of the iodinated immunoglobulin
surface molecules into the medium. The immunoglobulin molecules are
not released "free", but on fragments of plasma membrane. During this

period of time, there is no detectable release of H-2 alloantigen. I suggest that the most likely explanation is that microvilli, which are very dense on the surface of B lymphocytes (one of the major classes of lymphocytes in the spleen) are pinched off, and that immunoglobulin and many other proteins can migrate to the tips of the microvilli. I would suspect that these are proteins which are not deeply embedded in the membrane.

In contrast, I speculate that H-2 alloantigens are probably deeply embedded and for reasons that are not clear, cannot migrate to the tips of the microvilli, perhaps because of protein-protein interactions or because of an attachment to structures on the inside of the membrane.

The second point is that there is an inaccessibility of a significant proportion of H-2 alloantigen and a much smaller proportion of immunoglobulin molecules for interaction with ligands in the medium. This was a very puzzling finding to us. One way of measuring cell surface (as opposed to intracellular) immunoglobulin is to determine its ability to bind antibody in the medium. Thus when iodinated intact cells are reacted with antibodies to immunoglobulin, the cells lysed, and the antigen-antibody complexes precipitated with antibody to the antibody, the percentage of immunoglobulin molecules on the surface that are available to react with bivalent antibody can be quantified. In the case of the Burkitt lymphoma line (Daudi) 100 per cent of cell surface immunoglobulin can be bound by antibody. When the method was applied to mouse splenocytes, approximately 15 per cent of immunoglobulin molecules could not react with antibody to immunoglobulin. On the other hand, in the case of H-2 alloantigen on splenocytes, the figure was approximately one-third. In addition, the binding of anti-immunoglobulin was not temperature-dependent, whereas that of anti-H-2 alloantigens was markedly temperature dependent, that is, binding was much more effective at 37°C. than at 4°C. At this point we worried whether there might be a trivial explanation concerned with the methodology, which was very complex. Dr. Vitetta therefore performed experiments with papain, which gave strictly analogous results. Under conditions of papain digestion of intact cells in which 80 per cent of cell surface immunoglobulin was rendered non-antigenic, only 20 per cent of H-2 alloantigen was so digested. The two molecules in solution appeared to be equally susceptible to papain. Moreover, the same molecules that are not able to bind antibody are the ones that are not affected by papain. There seems therefore to be a peculiar inaccessibility of a percentage of H-2 alloantigen and a smaller percentage of immunoglobulin to interact with ligands. I cannot offer a unique interpretation. The data are consistent with the idea that there is some restriction on movement of membrane molecules, particularly those that may be deeply embedded in the membrane. Perhaps H-2 alloantigen exists in micropatches at the base of microvilli, so that some of the molecules in the inside of the patch are not readily available to interact with ligands. There are, of course, other explanations for these findings.

MARCHESI: Could this be a new membrane adjoining the cell surface?

UHR: Do you think some of these molecules may be internalized? That was excluded for H-2 alloantigens by iodinating cells, incubating, and observing whether the capacity to bind antibody changed with time of incubation. If there was constant internalization of the cell surface, one would expect the ability to bind antibody to decrease. In contrast, it increased modestly.

CUATRECASAS: Do the differences represent characteristics of single cells, or of different cell colonies?

UHR: H-2 alloantigens are on all nucleated cells; and in roughly equivalent amounts on different populations of lymphoid cells. In fact, they are particularly concentrated on small lymphocytes which makes up to 90 per cent of the population studied. Therefore one cannot explain 35 per cent of the H-2 alloantigens which are unavailable to ligands by postulating that they are on a cell type other than the small lymphocyte. I cannot comment on whether a portion of the molecules on all lymphocytes are unavailable to ligands or whether all the molecules on a subpopulation of small lymphocytes are unavailable.

MARCHESI: What is the evidence that there are any trans-membrane proteins in the lymphocyte membrane?

UHR: I don't think there is any evidence comparable to the evidence in the red blood cell. H-2 alloantigens are relatively insoluble. If they are digested on intact cells by proteolytic enzymes a soluble portion, approximately 90 per cent of the molecule is removed, and there remains an insoluble residue, which I believe Nathenson has shown to have a high content of hydrophobic residues. No sequencing has been done. Immunoglobulin in contrast, is a highly soluble molecule, at least as it appears in the circulation. No structural differences have been pinpointed as yet between immunoglobulin in the serum and immunoglobulin as it exists on the cell membrane.

MARCHESI: But there is no direct evidence that indicates that any part of the proteins actually goes down into the cell?

UHR: No.

RAFF: It should be pointed out that some of the things that Uhr has said are controversial. I don't think there is any convincing evidence that enables one to predict that H-2 penetrates through the membrane and that

immunoglobulin does not. It is also an oversimplification to say that immunoglobulin G is more readily mobile in the membrane than H-2. I
think there is some evidence that immunoglobulin in fact is not normally
concentrated at the tips of microvilli, because if you look at cells with a
monovalent anti-immunoglobulin conjugated to ferritin, you see an even
distribution over the villi. So there is no evidence for a natural flow of
immunoglobulin towards the villi. I think that the same is probably true
for H-2. Finally, it should be stated that there are conflicting data on
the turnover rates of both H-2 and immunoglobulin on lymphocytes, and
I do not think it is clear as yet that they are significantly different.

KAHAN: Jonathan suggests that there are two different classes of surface
molecule, one type being exemplified by immunoglobulin and the other type
by transmembrane histocompatibility antigens. I would suggest that a good-
ly portion of the histocompatibility antigen of lymphocytes behaves as if it
is in a superificial location, namely it can be liberated from cells by relatively mild treatment with chaotropic agents which are believed to only
disrupt water structure within the membrane. Although some antigen may
exist in a transmembrane location, there is certainly no direct or indirect
evidence to support this. Failure to find the more extrinsic H-2 molecules,
which are readily solubilized, shed into the medium in Jonathan's experiments
is worrisome. However, it should be pointed out that these findings are
not in real contradiction with our own which show recovery of HL-A antigens in the medium in which lymphoblastoid cells have been grown and in
the serum. The antigen in the medium is associated with membranes and
probably represents residua of dead cells. The antigen of the serum may
represent a digested form of the native material on the cell surface resulting from the activity of proteolytic enzymes on membrane fragments,
or may represent a special secretion process by a as yet unknown, cell
type other than the lymphocyte. It may be that the existence of circulating soluble antigen regulates secretion by lymphoid cells, and accounts
for the lack of active secretion in the relatively short time period in Uhr's
experiments.

MARCHESI: There is a little controversy brewing. We would like to hear
from Jerry Edelman.

EDELMAN: I don't want to get in the middle - but a different question.

MARCHESI: No, I want to know if you can suggest how immunoglobulin is
binding on the membrane. That may be relevant to the argument.

LERNER: I think that there is a concept here that is very important to
bring into focus since we also discussed it some years ago. It seems

obvious that if one can get a variance anywhere in biology, the immunoglobulin system provides an excellent tool. In this system there is a
situation of two compartments, that is antigenically similar molecules
which either are inserted into the membrane or secreted, depending on
the state of cell differentiation. We believe that in some cells both events
are taking place simultaneously, and on kinetic grounds I think this is
fairly solid. I believe that it is not necessary to make comparisons between molecules like H-2 and immunoglobulin. I think one can simply
ask how does the cell determine whether a molecule will be secreted or
inserted into the membrane when it is structurally really quite similar.
I think that maybe some of the people here will have some notion in this
regard. I really believe that these are the kind of comparisons which are
important. In other words, the cell is capable to utilize structurally similar molecules for quite different purposes, depending on the stage of the
growth cycle or the differentiation of the cell. I think that this is an important consideration to keep in mind.

MARCHESI: This comment is exclusively directed towards the immunoglobulin situation. There is no evidence in any other membrane system
that membrane proteins are being secreted.

REYNOLDS: That is not true. There is at least one other system- dop-
amine -beta- hydroxylase in chromaffin granules. These granules are the
secretory vesicles from the adrenal medulla. The enzyme exists as about
50 per cent soluble protein in the cytoplasm and is secreted into the plasma
with catecholamine release. The other 50 per cent of the enzyme is tightly associated with the membrane and can only be released by agents which
disrupt lipid bilayers, such as detergents. As yet no one has convincingly
demonstrated whether or not there is any structural difference between
the soluble and membrane-bound forms. It is intriguing to consider the
possibility that the membrane-bound species is identical but tightly associated with a specific number of membrane binding sites.

MARCHESI: Chemically it may not be the same molecule.

LERNER: If you will pinpoint what anchors the protein to a membrane
like the lymphocyte membrane, it is probably more profitable to look at
the anchor of a similar molecule rather than to look at the question with
two different molecules. We already have enough trouble with the same
molecule.

MARCHESI: That's why it may not be profitable to conduct these two arguments simultaneously.

EDELMAN: I would like to make a point about what has been said. It seems to me John (UHR) that you have to separate your observations from the particular interpretations - not that I take issue with either. What you have observed is a difference in susceptibili ty to a chemical reaction. In view of the fact that we don't know any details about H-2 structure we can't get too detailed. Forgetting that, and trying to answer Marchesi's question, there is perhaps something to be said about immun-oglobulin. If we assume that the immunoglobulin on the cell is the same as that in solution, then CH3 is probably the domain interacting with the cell. This probably has a fair number of polar and charged residues on the outside and so you will have a little difficulty sticking it into a medium with a dielectric constant of 2. This does raise an intriguing idea - that there are proteins in the bilayer that protrude and act as carriers for immunoglobulin which interacts with them rather than the membrane. I believe that Milstein has recently suggested that the F_c receptor may be the carrier. This is amusing and worth discussing.

The question which I had was directed to McConnell and Raff. What is the dimension of a microvillus. In particular, is the radi-us of curvature of the order of molecular dimensions? Because if it is, the proteins would not have much room to stick through the bilayer at the tip. Do you see ferritin label at the tip?

RAFF: Yes.

EDELMAN: How many angstroms?

MARCHESI: One hundred.

McCONNELL: We have made an interesting observation since we found that acetylcholinesterase can be selectively removed from red blood cells by transfer to phospholipid vesicles that have a diameter of about 200 Å. The enzyme remains fully active and there is no apparent fusion of the vesicle, so that it is possible to have active enzymes in very small ob-jects. With respect to some of the other points that were raised ....

BRANTON: Before you get off the subject, do you mean specifically that size vesicle and no other size vesicle?

McCONNELL: We have tried that size vesicle and it worked. I have read that some red blood cells contain globoside and that anti-globoside anti-body does not react to the cell. This is thus not a simple case where one has a cryptic antigenic substance on the cell surface which is in line with some of the arguments advanced by Uhr for a much more complex system.

MARCHESI: Yes, there is an analogous problem, and we will come to that during the discussion. However, we ought to finish with John Uhr on this point.

UHR: I don't think that the controversy that Raff's and Kahan's remarks would lead one to believe actually exists. First of all, the ideas that have emanated from the series of experiments that I mentioned are clearly highly speculative. The data are only indirectly related to the position and mobility of H-2 and immunoglobulin in the membrane. The fact is, there are marked operational differences between the behavior of H-2 and immunoglobulin on the surface of B lymphocytes, as well as the known biochemical differences between these two glycoproteins, and I firmly believe we should search for an explanation of these differences. Raff spoke about the presence of these two molecules on microvilli. I don't think this conclusion can be accepted unless scanning immuno-electron microscopy is performed.

        In regard to the structural signal which enables immunoglobulin to bind to the cell, I think one finding is clear: there is no extra piece of polypeptide chain which is over 15 - 20 amino acids in length which is covalently bound to membrane immunoglobulin and to the same class of secreted immunoglobulin.

MARCHESI: The point Edelman raises is interesting, and it focuses on the question of what is a receptor molecule. There must then be another molecule just like that which has the free mobility within the bilayer.

KARNOVSKY: Relative to what Uhr said, the possibility exists that the problem of accessibility could be related to the disposition of these molecules on the membrane. This is very controversial in terms of H-2. Some have claimed it is dispersed in the form of patches. I don't want to discuss it, because there is no assurance as to the validity of the methodology at the present time. But recently with regard to immunoglobulin, Drs. Abbas, Ault, Unanue and I have used ferritin labelled monovalent immunoglobulin. The pattern as seen in freeze-etching of mouse B cells, looks random to the naked eye, but if you do a morphometric analysis, you find that this pattern is actually highly non-random. There are micro-aggregates organized in an ill-defined reticulum with bare membrane in between. I think this might be of interest in terms of the partial inaccessability of antibodies to immunoglobulin. This non-randomness persists at 4° or 37° C.

BRANTON: Have there been experiments, for example, utilizing cross-reactions to determine what the immunoglobulin might be linked to on the surface of the membrane?

EDELMAN: Uhr is shaking his head with tears in his eyes.

UHR: No, but -

BRANTON: In other words, the results are all negative, but they have been done.

UHR: That is reasonable.

KENNEL: There is one small point that I would like to make. In isolating membrane immunoglobulins from all the lymphocytes we have looked at, we find as Uhr has found earlier, the IgM monomer type molecule. When this material is analyzed in a low percentage acrylamide gel in SDS, it shows a mobility that would indicate that it has a molecular size larger than one would expect. Whether this difference is due to the additional binding of SDS or to a differential shape of the molecule, it may nevertheless have some bearing on the mode of attachment to the membrane. If one is looking for a "handle" to hook into a membrane, this might be an indication of what it is. Like Uhr we have been unable to find any other molecules attached to the monomer structure.

MARCHESI: There is still a hope there will be some hydrophobic peptides that nobody has got as yet.

KENNEL: It will be virtually impossible to prove, because to isolate enough of this material to do sequence analysis is an incredible task. We are not working with red cells.

MARCHESI: Before we continue, I think it is important that we discuss the other class of molecules located at the membrane that contain residues that have receptor functions, have antigenic activity, and may or may not fall into this category of being exclusively on the outer edge of the membrane. These are glycolipids, and Dr. Cuatrecasas will introduce the subject.

CUATRECASAS: These are interesting molecules that we will be talking about. They are not macromolecules, yet they are accessible to the kinds of analysis we have been discussing this morning, and in many ways they are much simpler things to deal with in terms of chemistry. Manipulation may be much easier to study structure-function relations.
      I suspect that most of you do not know very much about glycolipids; they have certainly been very much neglected in membrane studies. I am talking about glyco-sphingo-lipids. I think I should spend a few minutes describing their chemistry.

I'll make this very simple, just to convey the basic aspects of their chemistry, but of course it is much more complicated. The basic structural unit of sphingolipids is the ceramide portion, which consists of a fatty acid attached to sphingosine. The latter is really a generic name. There are many kinds of sphingosines. The C-18 derivative is a prototype. Basically all we need is a long chain base having a 2-amino-1,3-diol derivative. However, in some cases there are no double bonds, or two double bonds, and there may be other hydroxyl groups. The important thing is the basic sphingosine-type structure, in which the amino group is esterified in this case with steric acid, which is the most commonly found fatty acid in the nervous system; but many others can be present here.

What determines the exact nature of the sphingolipid is the nature of the substituent at the hydroxyl of carbon one, which group is esterified at the hydroxyl. For example, here we have sphingomyelin, which has on the first carbon, phosphorylcholine. Other classes of glycosphingolipids include neutral sugar substituents, such as glucose, which is glucocerebroside; oligosaccharides of various types can be substituted. Another common derivative is sulfatide with a sulfate on the sugar. Another large and important class is gangliosides, which have an oligosaccharide containing sialic acid. The only thing which basically distinguishes a ganglioside from other sphingolipids is the presence of sialic acid.

What is known about the localization of glycolipids in cells? I think most people agree that they are found in membranes. The principal points of discussion are in the first place, what proportion of these molecules are present in the plasma membrane, and secondly, what is their orientation in the plasma membrane? Are they located so that the carbohydrate is inside, outside, or both? Clearly, they do not span the membrane, since the ceramide portion even in extended form cannot be larger than 30 Å. There has been much controversy, especially in the past, concerning the accessibility of these glycolipids to the exterior, even if they are indeed localized so that they are facing toward the outside of the membrane. What is their accessibility to react with molecules in the medium? The most common method for examining this question is simple chemical analysis. This, of course, can only tell about gross localization, if you are able to isolate a very clean fraction of cell membrane. It can tell you whether and what kind of glycolipid or spingolipid are present, but does not tell anything about orientation or dynamics of localization.

The second general approach is enzyme digestion. In the past, a great many studies using various hexoaminidases, galactosidases, and neuraminidases, have suggested that most glycolipids are not susceptible to digestion by those enzymes. However, in most of these studies, it has been a total membrane which has been examined, and one can only properly analyze the effects by looking selectively for changes in the plasma membrane. In fact, Murray has done interesting studies showing that if you look at the plasma membrane of normal and of transformed 3T3 cells digested with neuraminidase, approximately 50 per cent of the lipid-bound sialic acid can be removed, indicating that at least certain glycolipids are exposed to the medium.

Using antisera, Hakamori, who has probably done as much work as anyone in the area, described a nice study in which he looked at the ability of anti-globoside, antihematoside and antilactosylceramide antibodies to agglutinate erythrocytes. He found that antisera did not agglutinate adult erythrocytes very well at all, but did agglutinate fetal erythrocytes. But if the adult cells were digested with trypsin, they agglutinated very well, whereas the fetal cells did not change after such digestion. This is a basic observation which shows again that at least certain glycolipids (or a certain population of these) are accessible, and in addition, that there may be differences in masking in certain cells.

Another approach taken also by Hakamori in a very recent study is the method of labelling with galactose oxidase, followed by tritiated sodium borohydride. In this case he made the observation that globoside could be labelled in the intact cell. This method of course labels galac - tose and galactosamine which are present both in glycolipids and glycoproteins. He also found that globoside could be labelled both in adult and in fetal erythrocytes, and that trypsin treatment did not make very much difference. Again, the principal point is that these glycolipids were accessible to galactose oxidase, Why both types of cells were equally susceptible to the enzymes on the one hand, and there were differences in the reactions with antibody on the other hand, is unclear.

This brings us to the important question of the differences in the organization of these glycolipids within the membrane. Certainly measurement of agglutination is not a direct measure of binding to the glycolipid, since it is a much more complicated process. This raises questions which I think we will come back to later, in terms of the glycolipid organization, the capacity to change their orientation, the possible involvement of other membrane macromolecules which may be closely associated with glycolipids, and the relation of all these to membrane fluidity.

Not much is known about the function of these molecules. Generally in the past they have been thought, particularly in the nervous system, to be primarily structural elements in the membrane. Although this may be true for some of the simplest sphingolipids, I think there is reason to suspect that it may not be the case with the more complex glycosphingolipids particularly the gangliosides. If nothing else, because of the exquisite specificity some of them display in their carbohydrate composition, and secondly because some of the gangliosides are present in very small quantities in the plasma membrane, I would guess that they may have some very specific and functional roles other than simply structural ones.

In the past, perhaps the area in which special glycolipids have been implicated most has been in the processes of growth and transformation. The observations made by many people including Hakamori, Brady, Robbins, and many others, are that there is a parallel between the complexity of the glycolipids, particularly the gangliosides, and the rate of growth or the state of growth of the cell, so that cells growing very rapidly have much simpler gangliosides than cells which have stopped growing or are at confluency. On the other hand, cells which are transformed also have a much simpler ganglioside composition; the higher gangliosides are present in small quantity. But the existence of the parallel is really all that can be said, and there are exceptions.

We are performing studies with cholera toxin, an exotoxin from virulent cholera, which provides an interesting tool for studying glycolipids, especially a certain class of gangliosides. Cholera toxin, a molecule of about 100,000 daltons molecular weight, has the property of stimulating adenyl cyclase in every tissue of mammalian origin which has been examined (providing adenyl cyclase is present); it is ubiquitous in its action.

We have labelled cholera toxin with ^{125}I to high specific activity, about 2 Ci per millimole, and the molecule remains quite intact biologically. In fact, you can double the amount of label, and it still retains full biological activity. The radioactive molecule binds very well to cells, to a variety of cells, and to a variety of plasma membranes. It is easy to bind 60 per cent of the radioactivity in the medium at a concentration of less than 10^{-10} M. There is a very sharp competition by native cholera toxin; the displacement is nearly complete, and there is almost no non-specific binding. This is really extraordinarily tight binding - there is no hormone, for example, which binds anywhere the way this molecule does. The binding is very fast, consistent with the extraordinary avidity of the binding. For example, at 10^{-10}M, even at 4°C. the binding is complete in about ten minutes, perhaps sooner depending on the concentration of membranes. It is possible to saturate the binding in most cells. Certainly it is possible to do this within the concentration range of physiological effects.

In a fat cell the maximum number of molecules which can bind is really very small, about 10,000 per cell, which for example, is very close to the number of insulin receptors. This small number is not true for all cells, and we will come back to this later. There are some cells that bind a great deal more, 100,000, and there are cells that bind much less, under 1,000. The rate of dissociation of the molecule is very slow, consistent with the high affinity.

Now, what is the toxin binding to? It is binding to a specific glycolipid. We found that a number of neutral glycosphingolipids do not bind significantly. As we increase to greater complexity of glycolipids and the gangliosides, there is no binding. More complex lipids such as GM3 shows virtually no binding. Now simply by putting a galactose residue in the terminal position now affords a molecule GM1 which is extraordinarily specific in its ability to bind to the toxin and prevent the toxin's attachment to cells. Adding a sialic acid to GM1 causes a loss of the binding ability. When the GD-1-A compound, which is GM1 attached to a single sialic acid is digested of this acid residue, then binding is restored. So GM 1 has extraordinary specificity and the binding is very tight. Figure 10 shows that toxin pre-incubated with GM 1 is incapable of binding to cells. We can inhibit binding at very low concentrations of the ganglioside. The fall in the binding parallels very closely the inhibition of the biological activity of cholera toxin. It is clear that within the proper range of concentration, the biological effect is correlated very closely with the number of molecules of toxin which bind to the cell.

If instead of putting GM1 ganglioside in the medium, or pre-incubating it with toxin, we pre-incubate the cells themselves with GM 1 and then wash them thoroughly before measuring binding, the treated cells bind very much larger quantities of the toxin, suggesting that there is a spontaneous incorporation of the ganglioside into the cell membrane. The incorporated molecules are subsequently able to bind cholera toxin. The binding properties of these "new" sites are very similar, if not identical, to those which are present normally.

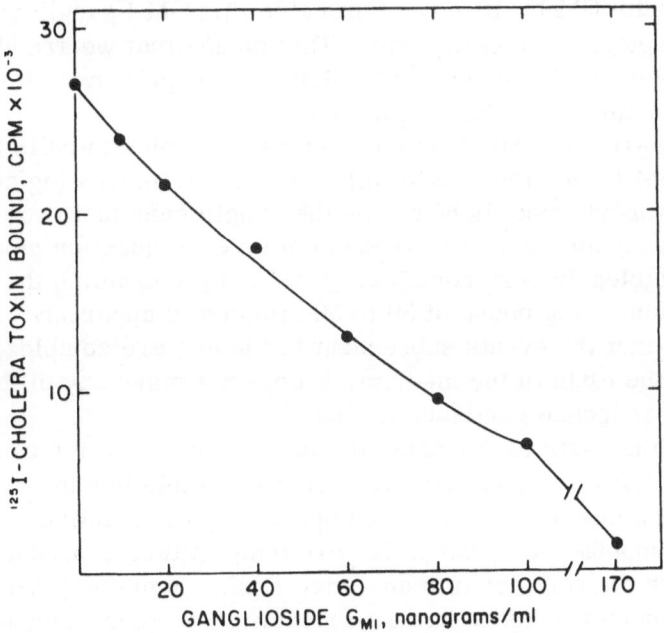

Fig. 10. Inhibition of toxin binding by pre-incubation with GM1.

Monosialoganglioside GM 1 is a beautiful amphipathic molecule with two very different chemical domains - a carbohydrate region which is very hydrophilic and a ceramide portion which is highly hydrophobic. It is suggestive of the hypothesis that the latter portion is anchored into the membrane and the polar region is exposed to the medium where it must interact with cholera toxin. It must be accessible to a very large molecule (molecular weight approximately 100,000 daltons). Ganglioside GM 1 appears to insert into the membrane very well. The molecules which are put into the membrane are very stable, and they can stay there for days. There may be virtually no loss in the enhanced binding up to five or six days. In tissue cells it is incredible how tenaciously it will remain bound to the cells, even following cell division.

We can say that this is all very interesting, that this molecule can bind to the toxin, that it can go into membranes and that when artificially added to membranes it can bind the toxin. But how do we know whether this binding is related to the natural receptor binding? In other words can the binding of the toxin to exogenously added ganglioside result in a biological response? Is it truly a receptor, or is it something else?

The answer is that it does modify the response. Under certain conditions, if one manipulates the system correctly, it is possible to observe no biological response in untreated cells, while there is a maximal response in cells which have been treated with GM 1. The difference in activity is accounted for by an increase in the amount of binding in the cells treated with GM 1. It is very probable that GM 1 ganglioside is the chemical receptor for cholera toxin. This means that we are able to reconstitute a biological system, a modulator - receptor system in the plasma membrane that involves a key regulatory enzyme.

We are now extending these kinds of studies to cells that have virtually no GM 1. Another issue which is equally interesting and which we can now properly ask, is how does the ganglioside-toxin complex affect the enzyme adenylate cyclase? This is not an easy question since the processes are clearly very complex. If we simply examine the time course of action, a lag phase of 30 to 60 minutes is apparent, and this alone tells us that the events subsequent to binding are complex. If one washes away the toxin in the medium, it does not make any difference; the biological response continues unabated.

On the basis of our data, the mechanism of action can be explained in the following way: The first step is simple binding of the cholera toxin to the ganglioside, forming a complex which is inactive. Something then happens spontaneously which is very temperature dependent and which converts the complex into an active form. Something happens during the lag phase such that there is a re-orientation or re-organization within the two-dimensional framework of the membrane, such that the "active" complex becomes associated with and perturbs adenyl cyclase. At present we visualize a direct physical interaction and modulation of activity. A very useful structural analog of cholera toxin is called **choleragenoid**, which is a subunit of the protein and can combine with GM 1, but is not biologically active. It cross-reacts immunologically, and is able to bind with precisely the same properties as cholera toxin, thus it is a competitive anatagonist. The derivative is able to form the original complex, but it cannot undergo the subsequent transition. There is much more to say about this transition.

One of the interesting properties relates to the nature of the dissociation of cell-bound toxin. The extent but not the rate of the dissociation changes with time. If binding is done for 3 minutes at 24° C., the toxin can dissociate spontaneously almost completely. However, if it is bound for 10 minutes, only 70 per cent can dissociate and after 20 minutes at 37° C only 50 per cent can subsequently dissociate. There are other kinds of experiments of this type which indicate that something is changing with time, and that the molecule is not able to come off the same way that it did previously. Figure 11 shows some of the important features concerning the structure of cholera toxin, which is indeed a very interesting molecule. The molecule consists of two basic portions, which we call the "binding" and the "active" subunits. The binding subunit binds ganglioside and corresponds to the choleragenoid molecule that I referred to earlier. It consists of an 8,000 daltons molecular weight subunits, which appear at present to be identical, but this is not absolutely certain yet.

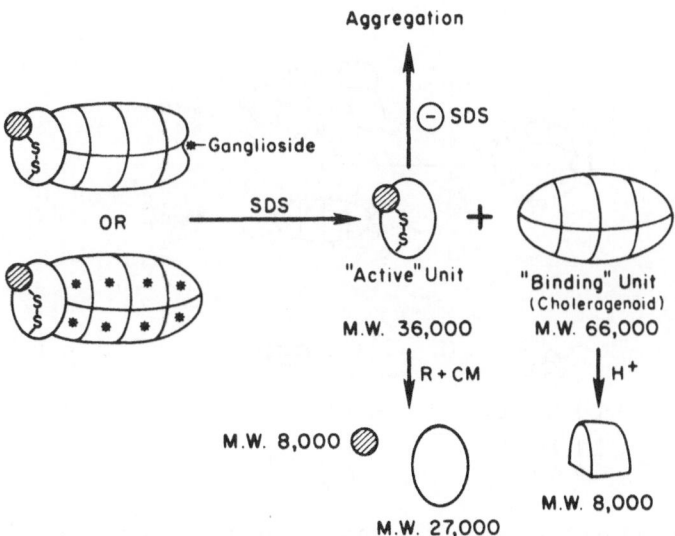

Fig. 11. Postulated molecular structure of cholera toxin.

There are two possibilities suggested for the binding, involving one ganglioside per choleragenoid or one per 8,000 molecular weight subunit. We favor the latter since there is evidence for multivalency, and we feel that there must be at least two binding sites for gangliosides in the native molecule. In SDS it is possible to dissociate the active and the binding subunits. Interestingly, the binding unit, the choleragenoid, does not dissociate in SDS unless it is heated. This is thus another example of a non-membrane protein which is not readily dissociated by SDS. The active unit alone does not bind, and is devoid of biologic activity. Another feature about this subunit is that in the absence of SDS it aggregates markedly. It is virtually impossible to maintain it in solution without SDS or other detergents. There are some features of this molecule which suggest that it has many properties of an integral membrane protein. Figure 12 presents a hypothesis which is based on these structural features and on some experimental evidence. The intact toxin may bind to the ganglioside through the choleragenoid, and during the subsequent transition, there may be dissociation of the molecule, such that the active, hydrophobic subunit becomes incorporated with membrane. Here it becomes a "permanent" component of this membrane. Drs. Vann Bennett and Susan Craig in my laboratory have some data suggesting a differential rate of dissociation between the active and binding subunits, suggesting that the active unit may bind much more strongly than the other. This is the portion of the molecule that once it gets into the membrane, does not come off. It appears to behave like an intrinsic protein of that membrane, and to remove it one must dissolve the entire membrane with detergents. This

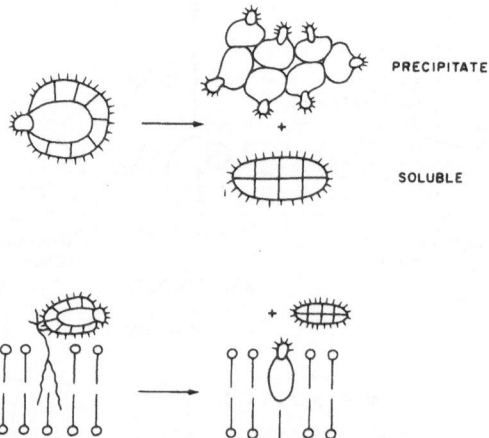

Fig. 12. Hypothetical incorporation of hydrophobic subunit of cholera toxin as intrinsic membrane protein.

is the portion of the molecule that we at present believe associates directly with adenyl cyclase. Very recently, Bennet in our laboratory has made affinity columns containing the active unit. These columns can adsorb solubilized adenyl cyclase very substantially. Adenyl cyclase appears to stick quite strongly.

MARCHESI: This is a fascinating idea that exogenous molecules which are not membrane proteins become incorporated in the membrane. As far as I know, no one has ever proposed this or provided evidence for it. One thing that strikes me is how is it that the active unit does not pull out the ganglioside with it? Is there a certain amount that comes off the active complex? Why doesn't a certain percentage of the ganglioside come out and form a micelle?

CUATRECASAS: It is really curious that it does not appear to happen, and suggests that the forces maintaining the ganglioside in the membrane are very great indeed. In fact this is another reason why we believe that these glycolipids are associated with very specific macromolecules in the membrane. By the way, the fact that insertion into the membrane happens spontaneously does not mean that it is entirely passive. The process has a sharp dependence on temperature, and there is also a maximum amount of ganglioside which can enter. There is much evidence suggesting that there are other molecules in the membrane with which the ganglioside is specifically interacting, and that under certain circumstances, these may associate with each other to form macropatches and even caps on the cell surface.

MARCHESI: Concerning the time lag - if you add exogenous ganglioside to the cell, is there a correlation with the amount of ganglioside you can add?

CUATRECASAS: No, we cannot shorten the lag period beyond a certain point. We can lengthen it, but we cannot shorten it.

EDELMAN: Do you have any idea how big that hydrophobic central core might be? Is it as you depict it, which would suggest 30 angstroms.

CUATRECASAS: It is at a maximum around 30 angstroms, if it were an extended chain.

EDELMAN: Is it true of other glycolipids that they seem to be anchored as well, not just this specific set, but the general family of glycolipids. So they also fit pretty tightly in the membrane, or are there classes that come out more easily?

CUATRECASAS: I do not know of any other class that has been looked at quite like this.

HAKAMORI:: When we grow NIL cells in a medium containing globosides which are isotopically labelled, we find that globosides very quickly incorporate almost exclusively into the membrane, and markedly change the growth behavior of the cells. Extended pre-replicative phases make cell growth slow down. We are now studying the effect of exogenous globosides on various physiological properties of cells, such as cyclic AMP level and transport activity.

EDELMAN: Do they compete with each other?

HAKAMORI: This particular cell does contain globosides originally, and its concentration increases to a critical point. I do not know if there are structural changes.

EDELMAN: If you add excessive amounts of different glycolipids, would it finally saturate out so that it would compete against detergent?

CUATRECASAS: These are interesting experiments which have not been performed yet. This would be easy to do because of the availability of labelled gangliosides. I might mention another point which is very pertinent. Certain cells have a very large number of toxin binding sites, such as 100,000, while the number of adenylate cyclase molecules (presumably), and certainly the number of toxin molecules needed to get a maximal response is very low, for example 4,000. The interesting thing is that it appears that all 100,000 toxin sites (gangliosides) are equivalent with respect to their potential ability to modify and interact with the cyclase. They are therefore probably not associated with each other to begin with. All of these gangliosides appear to be equivalent with respect to their ability to substitute or associate with or modify the enzyme, generally suggesting that there must be some redistribution or change after binding.

HAKAMORI: I think that the exogenously added and incorporated glycolipid into cell membranes behaves like an original membrane component, and that antibodies directed against this particular carbohydrate lyse the cell in the presence of complement. Blood group O cells can be converted to blood group A cells by exogenous A-glycolipid.

KATZEN: I have two questions. Presumably all cells are stimulatable by cholera toxin. Thus how do you obtain a cell that is not stimulatable by the toxin so that you can add GM 1 exogenously? You are at a maximum level already.

CUATRECASAS: This can be done, and has been done, under conditions of a very small concentration of cholera toxin and very dilute cells, such that you bind only a small proportion of the toxin and occupy only a small proportion of the available sites. If a cell under these same conditions has ten times the number of receptors, for example, that cell will bind more of the cholera toxin present in the medium. This is a bimolecular reaction. The one thing that counts is the amount of toxin bound to receptors on the cell, and this can be increased in two ways: one, by adding more toxin to the medium and secondly, by adding more receptor to the cell. In the case of the toxin, you can achieve the same effect by manipulating either component. There are other cells, transformed cells, K/BALB cells, which have a very minimum response to the cyclase, because the content of GM 1 is negligible. In those cells it is much easier.

KATZEN: Since you have indicated that only a small percentage of the receptors need to be bound to give complete activation, so that the remainder are excess or spare, and since the adenyl cyclase units on the intact membrane are already presumably saturated with excess intact GM 1 receptors, it is puzzling that addition of exogenous GM 1 is both capable of binding to membrane sites linked to adenyl cyclase as additional receptors, and of activating sites that are already completely "activatable". Wouldn't you first have to get rid of endogenous GM 1?

CUATRECASAS: In fat cells the number of binding sites available happens to correspond very closely with those required for a maximal response, which I believe is fortuitous. The experiment is thus not so difficult. It should be even easier in cells in tissue culture, which we are now examining, that have virtually no GM 1, no binding, and no response in the absence of added gangliosides.

KATZEN: My other question relates to our own studies with the insulin receptor, and the studies by Steiner, which indicate a close correspondence between the degradation of the polypeptide hormone and its binding to the intact cell. In our case, we find a considerable amount of insulin degrading activity associated with crude soluble receptor extracts of the cell membrane, as does Steiner with the intact cell. In the case of the cholera toxin peptide, I was wondering how you might eliminate this type of degradation that might be occurring, which might obscure or cause a misinterpretation of your results with regard to the release of [125] I - labelled peptides. I refer to your measurements of the reversible binding and release of labelled toxin.

CUATRECASAS: The toxin is virtually unaffected by trypsin, and it is extraordinarily resistant to all known proteases. I don't know what findings you have in mind that might be explained by proteolysis. When I refer to specific qualitative subunit dissociation, I refer to identification in SDS disc gels. We have not seen anything that might suggest hydrolytic breakdown.

KATZEN: Could it be that there might be degradation related somehow to the debinding of the cholera toxin that you see, as well as to the very delayed debinding that you showed for the choleragenoid? I was just wondering if other mechanisms might be involved than simply the binding or activity sites on the toxin structures, in terms of the reversibility of binding which you measure.

CUATRECASAS: I'm sure there are a lot of factors, many of which we have not even considered yet.

LERNER: I think an issue was raised which we are going to see coming up over and over again: that is, what is a receptor site and what is a binding site. I think that not everything that binds to the cell does so via a receptor. Even selective displacement by a like molecule does not ensure that the interaction is biologically relevant. I wonder if it would not be worthwhile to reserve the notion of receptor for those molecules on the cell surface which bind another molecule in such a way as to alter cellular gene expression. This is not to neglect other molecules which may be extremely interesting probes of membrane structure. Otherwise, we may have more "receptors" than we have surface. Facetiously, one could speak about the "glass receptor" that makes a cell stick to glass. I don't know if you agree.

SESSION II

SOLUBILIZATION AND EXPRESSION OF
SURFACE MOLECULES

Extraction techniques - Membrane sugars and glycoproteins
of normal and malignant cells - Cell surface lactoperoxidase
and galactose oxidase labeling methods - Fusion dynamics of
cell membranes - Chromaffin granules - Soluble cell surface
derivatives - Molecular structure of HL-A antigens –
β_2 microglobulin - Mechanism of KCl solubilization.

SESSION II

SOLUBILIZATION AND EXPRESSION OF CELL SURFACE COMPONENTS

REISFELD: I would like to say a few words regarding some of the critical problems which face the investigator who seeks to solubilize a plasma cell membrane component while attempting to maintain its biologic reactivity. First, it is necessary to assume that such components can indeed be isolated independent of membrane structure with the solubilized moiety being in a form very similar to that originally present on the membrane. Once this assumption proves to be correct, the investigator who seeks to elucidate the chemical properties and molecular structure of such a soluble membrane component faces additional problems. Foremost among these is the difficulty in obtaining sufficient amounts of adequately purified materials to accomplish this task. This is largely attributable to the lack of highly selective solubilization procedures presently available which effectively render membrane components soluble. This is, of course, not surprising if one realizes the intricate and complex nature of the plasma cell membrane which was so well demonstrated during this morning's session. To isolate essentially hydrophobic components in soluble, highly purified form thus represents a considerable challenge.

The basic rationale for seeking soluble membrane components is, of course, to gain a better understanding of this complex cell surface at the molecular level and to utilize such isolated and chemically characterized components to correlate their chemical structure with their biological function. Prior to any attempt of isolating soluble cell membrane components, it is essential to probe the cell surface to determine their distribution in order to develop optimal conditions for their solubilization. Vic Ginsburg will now outline an approach which utilizes sensitive immunologic methods to assess carbohydrates and their distribution on the cell surface.

GINSBURG: I have written here on the blackboard the eleven sugars that occur in surface membranes. We are interested in certain sugar sequence that occur on cell membranes and we are trying to develop immunological methods to detect these sequences. Many of the sugar sequences that occur in glycolipids and glycoproteins are also found in the free oligosaccharides of human milk (Ginsburg, Advances in Enzymology, Vol. 36, ed. A. Meister John Wiley & Sons, New York, 1972, p. 131). We* have been isolating and characterizing these oligosaccharides for use in immunologic studies on the carbohydrates of cell surface. They are much easier to isolate than their counterparts in membranes and can be obtained in relatively large

---

* This work was carried out in collaboration with Dr. David A. Zopf.

amounts (A. Kobata, Methods in Enzymology, Vol. 28, ed. V. Ginsburg, Academic Press, New York, 1972, p. 262). The immunologic studies include: (a) the production of synthetic antigens by coupling oligosaccharides to carriers; (b) hapten inhibition of antibodies directed against c e l l surface carbohydrates; and (c) radioimmune assays for specific glycolipids.

The oligosaccharides are coupled to polylysine (M. W. 70,000) by a modification of a mixed anhydride procedure (Arakatsu, et al., J. Immunol. 97 (1966) 858). The coupled products, containing from 20 to 80 sugar residues per molecule of polymer, are effect ive antigens capable of precipitating specific antibodies. For example, the pentasaccharide lacto - N-fucopentaose II (LNF-II), with the structure Galβ1-3GlcNAcβ1-3Galβ1-4Glc, when coupled to polylysine is more

$\qquad$ 4

$\qquad$ Fucα1

effective on a weight basis than Le^a-active soluble human blood group substance in precipitating antibody directed against the Le^a blood group antigen as shown in Figure 13.

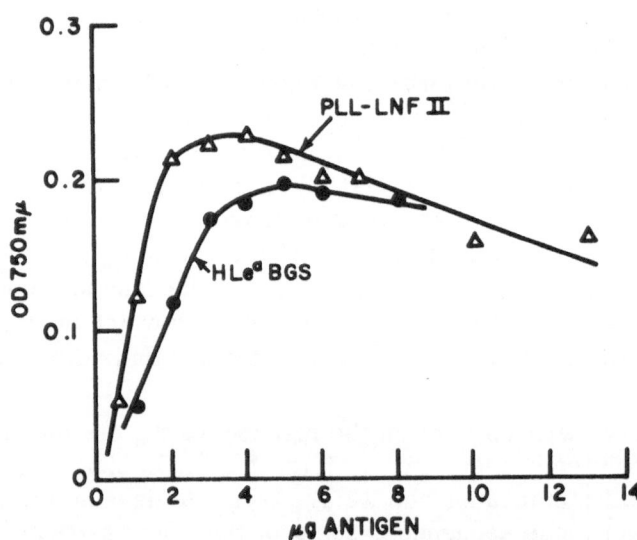

Fig. 13. Precipitin reaction of goat anti-Le^a serum (kindly supplied by Dr. Donald Marcus) with polylysine-LNF II conjugate (PLL-LNF II) and soluble human Le^a blood group substance (HLe^{a}BGS). The quantities of antigen shown were added to 50 μl goat serum in a total volume of 150 μl in isotonic NaCl buffered at pH 7.4, incubated 30 minutes at 37°C, then overnight at 4°C. Immune precipitates were collected by centrifucation at 10,000 x g, washed twice with buffered saline, and assayed for anti - body protein (Lowry, et al., J. Biol. Chem. 193 (1951) 265).

The antibody reacts specifically with LNF-II as shown by the inhibition of the precipitin reaction by oligosaccharides as shown in Figure 14.

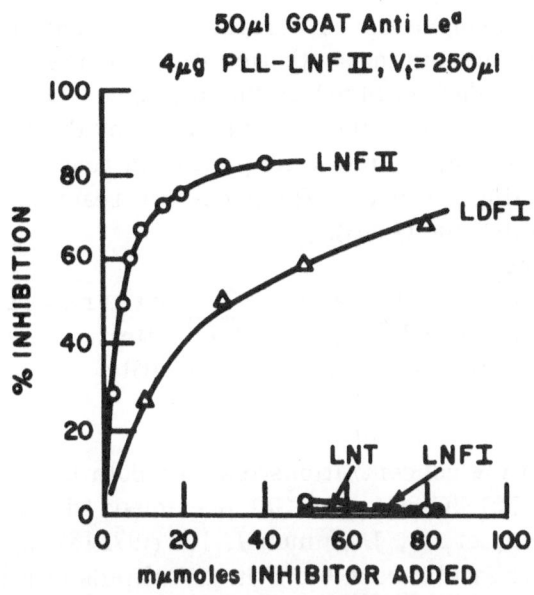

Fig. 14. Inhibition of precipitin reaction of goat anti-Le^a serum with poly-lysine-LNF II conjugate by milk oligosaccharides lacto-N-tetraose (LNT), lacto-N-fucopentaose I (LNF I), lacto-N-fucopentaose II (LNF II), and lacto-N-difucohexaose I (LDF I). Inhibitors were added to 50 μl serum in the quantities shown in a total volume of 210 μl in buffered saline and incubated at 37°C for 15 minutes; 4 μg of antigen was then added in 40 μl buffered saline and the final mixture incubated for 30 minutes at 37°C, then overnight at 4°C. Immune precipitates were measured as given in Figure 1.

LNF-II is the most effective inhibitor. The hexasaccharide, lacto-N-difucohexaose I (LDF-1), with the structure Fuc α1-2Galβ1-3GlcNAc β-3Galβ1-4Glc, also inhibits while lacto-N-fucopen-
                    4
        Fucα1
taose 1 (LNF-1), Fucα1-2Gal β1-3GlcNAc β1-3Gal β1-4Glc, and lacto-N-tetra-ose (LNT), Gal β1-3GlcNAc β1-4Glc, are inactive at the concentrations tested. Similarly, the polylysine conjugate of LDF-1 precipitates antibody directed against the Le^b blood group antigen and the precipitin reaction is specifically inhibited by LDF-1.

The conjugates were used to test the specificity of a human anti-T cold agglutinin (called "Ma") obtained from a patient with chronic autoimmune hemolytic anemia (Feizi, et al., J. Immunol. 106 (1971) 1578). The agglutinin specifically reacts with the polylysine conjugate of lacto- N-neohexaose (LNnH),

Gal β1-4GlcNAc β1

Gal β1-4GlcNAcα1-3Gal β1-4Glc. The maximum amount of antibody precipitable by the LNnH-containing synthetic antigen is the same as the amount precipitable by soluble human blood group substance "Og" as measured by Dr. Ten Feizi. Absorption of the serum fraction by the LNnH polylysine conjugate at the equivalence point of the precipitin reaction removes 95% of its hemagglutinating activity. The precipitin reaction is strongly inhibited by lacto-N-hexaose (LNH),

Galβ1-4GlcNac β1$_6$

Gal β1-3GlcNac β1-3Gal β1-4Glc; lacto-N-neotetraose (LNnT), Gal β1-4GlcNacβ1-3Gal β1-4Glc, is a weak inhibitor. Lacto-N-fucopentaose III (LNF-III), Gal β1-4GlcNAc β1-3Gal β1-4Glc, LNT and lactose

3

Fucα1

(Lac) are inactive at the concentrations tested. Both LNH and LNnH contain the sequence Gal β1-4GlcNAcβ1-6Gal previously shown to inhibit the anti-I of "Ma" (Feizi, et al., J. Immunol. 106 (1971 1678). As the LNnH structure has recently been found in glycolipids of human tissue (Wiegandt, Hoppe-Seyler's Z. Physiol. Chem. 354 (1973) 1049), it is possible that this sugar, or at least its distal part, is the natural red cell target of the agglutinin in vivo (the branched portion of LNnH occurs in glycolipids of human red cells (Hakamori, et al., Biochem. Biophys. Commun. 49 (1972) 1061).

In preliminary studies to develop sensitive methods for immune detection of specific carbohydrate sequences, we have tested the goat anti-Le[a] and anti-Le[b] sera described above in a radioimmune assay using nitrocellulose membranes (Gershman, et al., Prostaglandins 7 (1972) 407). Oligosaccharides from milk are easily labeled by reduction with sodium borotritide, converting the glucose at their reducing ends to sorbitol and the use of these derivatives in equilibrium dialysis experiments has already been reported (Zopf, et al., in E. Y. C. Lee (ed.), Biology and Chemistry of Eucaryotic Cell Surfaces, Acadamic Press, New York, 1974). Both the anti-Le[a] and anti-Le[b] sera bind the appropriate labeled sugar in the nitrocellulose membrane assay. The labeled sugar is displaced by homologous unlabeled sugars or glycolipids. Thus, if antibodies with high binding constants can be produced, a sensitive radioimmune assay for specific glycolipids is possible. Administration of polylysine oligosaccharide conjugates dispersed in complete Freund's adjuvant produces little antibody response in goats and rabbits. However, preliminary results of immunization with the synthetic antigens complexed with succinylated hemocyanin appear more promising and indicate that these materials may enable production of antibodies to chemically-defined carbohydrate antigens of cell surfaces.

REISFELD: How can these antibodies be used in the study of cell surface carbohydrates.

GINSBURG: They will enable us to measure levels of specific sugar sequences in membranes. We are particularly interested in measuring certain fucose-containing oligosaccharides that Dr. Hakamori finds accumulate in adenocarcinomas and not in normal tissue.

REISFELD: Do you feel that these sensitive probes will clarify any of the issues raised this morning?

GINSBURG: I think that antibodies whose specificity for sugar residues is rigidly established will yield much more information about the complex carbohydrates of cell surfaces than the lectins which are so widely used today.

HAKAMORI: The use of antibodies against pure carbohydrate is an extremely important approach for various cell biological studies as compared to the use of lections--of course, lectins are important, but lectins have a number of binding sites and the interaction of cells with lectins is complex, causing various secondary reactions. It is also impossible to obtain readily "monovalent lectins". In contrast monovalent antibody is relatively easy to prepare. Very similar studies which we are doing also utilize the glycolipid derivatives, oxidative ozonolysis products of sphingoglycolipids coupled to many different polymers which in turn enables us to prepare a monospecific antibody. It will be very interesting to compare your antibodies and our preparation.

REISFELD: In answer to your comment, there are a number of people here who have made extensive use of lectins and are using lectin absorbents. Would you think that the type of approach outlined by Ginsburg and you is to be preferred over the "lectin approach"?

HAKAMORI: Oh, yes, lectin is a lousy material--I am sorry to say that-- it is multivalent and thus a number of surface substances are reactive with it. Thus, if one reacts cells, this complicates matters. Also, the specificity of lectin leaves much to be desired as it is rather broad and really not very well known.

MARCHESI: I just want to ask a few questions. Why do you think that glucose is on glycolipids exclusively and manose is not on them? Is there any reason for that?

GINSBURG: Yes.

MARCHESI: What do you make of all these studies showing the glycosyl transferases on cell surfaces? Is it meaningful, does it modify all these sugars?

GINSBURG: Of course, lectin is very easy to obtain, purify, but as compared to the use of pure monospecific antibodies against carbohydrates I believe that it has less value.

LERNER: I think that it would be fair to say that although lectins may be a bag of worms, so are some preparations of antibody.

JEANLOZ: I would like to take exception to the absence of glucose in glycoproteins. In addition to have been found in structural (collagen-type) proteins, glucose has been found in glycoproteins isolated from its aorta wall. The problem of identification of glucose in glycoproteins is generally complicated by the use of Sephadex resins in the isolation procedures. The resins continually leak glucose polymers which contaminate the glycoproteins isolated. However, after isolating the TA3 cell glycoproteins without use of Sephadex columns, we could still find glucose among the carbohydrate components.

EDELMAN: Dr. Hakamori, I have this to say about both subjects: they are like love--none of it is bad, and you take what you can get.

REISFELD: That seems to settle the issue fairly well.

GINSBURG: You are quite right, I was talking in generalities. Glucose is found in two proteins, in basement membrane and collagen. They are structural and not functional like the glycoproteins we are talking about in membranes. The fact that glucose is found there, possibly in that particular glycoprotein, does not have a particular function--it aligns--

EDELMAN: I would like to direct a question to Dr. Ginsburg and Dr. Hakamori in relationship to the comments about the glycolipid receptors that Pedro Cuatrecasas discussed this morning. Is there any case in which a sugar recognizes another sugar in the same way that a protein recognizes the sugar as a receptor? In other words, is there specificity in sugar-sugar interactions?

KATZEN: I don't know if I have an answer to this, but it may be related to these questions. We have tried to overcome the problems of multivalency of Con A and are looking for its receptors on adipocytes. We do this by coupling purified Con A to Sepharose which only exposes a specific monovalent surface of the Con A. We have been able to bind this (as we have done with insulin) to intact fat cells in what we call "a buoyant density assay" wherein the fat cells float or sink depending on their binding to the Sepharose beads. We have demonstrated a very remarkable specificity of binding by measuring the inhibition of binding by exogenous sugars. We have synthesized many different analogues of various sugars and we have seen a remarkable specificity of inhibition by mannosides of the binding. But relevant to Dr. Edelman's question, a few of these sugar derivatives appear to be acting on the surface of the cell rather than with the Con A, as if there might be a sugar-sugar interaction, and we are quite intrigued by this at the present time. However, it is more likely that we are actually seeing a sugar-membrane protein interaction in addition to what we more frequently see as a sugar-Con A interaction.

GINSBURG: That doesn't seem likely.

KATZEN: We agree that the sugar-protein interactions are the most likely.

HAKAMORI: I should point out that this sugar-sugar interaction is not an old question, as was already discussed by Dr. Jeanloz this morning. About ten years ago, many people discussed interactions between hydroxyl groups of carbohydrates through hydrogen bonds; such interaction could be the basis of intercellular interaction also.

REISFELD: Undoubtedly interesting questions were raised by Dr. Ginsburg's presentation and some may have been answered. Dr. Jeanloz is now going to tell us more about some evidence indicating that glycoproteins are major components on the surfaces of tumor cells and that they have biologic importance.

JEANLOZ: I would like to discuss briefly some of the substances that we have isolated from the surface of two sublines derived from the TA3 cell. The TA3 mammary adenocarcinoma arose spontaneously in the strain A/ HeLa mouse, in Dr. Hauschka's laboratory at Roswell Park Institute many years ago. After being adapted to the ascites form and about 200 transfers, it was found to have lost strain specificity and the cells could be transferred even into the rat.

Around 1952, Gasic and Gasic demonstrated the presence of a heavy coat of glycoprotein around the TA3 cell and the decrease of pulmonary metastasis after removal of sialic acid from the cell by sialidase. For this reason, Dr. John Lodington in my laboratory, in collaboration with Dr. Barbara Sanford, started his work on the isolation of the sialic acid-containing substances at the surface of the cell and their characterization. In preliminary experiments, published in 1970, we were able to show that trypsin or pronase liberated more than 60% of the bound sialic acid as glycoproteins or glycopeptides. Trypsin was used at a concentration of 18 µg/ml and at 4° and, under these conditions, viability of the cells remained very high. Attempts to extract similar material with chelating agents, such as citrate or EDTA or, more recently, with lithium diiodosalicylate resulted in far less bound sialic acid being released as glycopeptides and in a marked decrease of the viability of the cells. Furthermore, the composition of the material extracted with the salts was very different from that extracted by proteolytic enzymes. We may conclude that the salts extract glycoproteins that are rather weakly bound to the surface of the cell, whereas proteolytic enzymes remove the external portion of glycoproteins embedded in the plasma membrane.

If we use neuraminidase, we can remove roughly 80 percent of the total amount of sialic acid present in the cell, and probably more than 90 percent of that present at the surface of the cell. If we compare the actions of trypsin and papain at similar concentration (70-90 µg/ml) for 40 minutes at 4°, we observe that all the bound sialic acid has been liberated as glycoproteins or glycopeptides. As the usual concentration of trypsin of 18 µg/ml, only one half of this amount is obtained. With pronase, we had an excellent release of peptides, but too much degraded for further convenient purification. I will discuss later the differences in composition between the materials extracted by trypsin and by papain.

About two years ago we were fortunate to learn that samples of the original species-specific TA3 cell had been kept in the laboratory of Dr. Klein at the Karolinska Institute in Stockholm. The non-strain-specific subline was called TA3-Ha and the strain-specific TA3-St. In a study made by Dr. Sanford we reported that both sublines possess approximately the same number of H-2 antigens at the surface of the cell, but that not all H-2 antigens are exposed at the surface of the TA3-Ha cell. Treatment with trypsin of the TA3-St cell released less sialic acid bound to protein, although the removal was as extensive as with the TA3-Ha subline. Thus there is far more sialic acid present at the surface of the unspecific line. If we calculate the density of the sialic acid residues at the surface of both type of cells, we find that the strain-specific Ha subline has about 2 to 3 times the density of the non-strain-specific St subline. The difference between both sublines became far more apparent when we purified the sialic acid-containing glycoproteins, on the basis of molecular weight, by passage through columns of P-4 and P-30 resins. At the surface of the TA3-Ha cell, Dr. Lodington found a high-molecular-weight glycoprotein, removed by trypsin, which contains about one third of the total surface sialic acid, and about one tenth of the total protein and carbohydrate material removed from the cell. This material was practically absent at the surface of the non-strain-specific St cell, less than one percent being detected. A similar, high-molecular-weight glycoprotein can be obtained from the material obtained by papain treatment of the Ha subline, none from its St subline. The carbohydrate compositions of the glycoproteins obtained by trypsin and papain are slightly different, which may be explained by microheterogeneity of the carbohydrate chains. The main chemical characteristic of the high-molecular-weight glycoprotein is the very high proportion of serine and threonine residues, about two-third of the total amino acid residues, the remaining being composed mostly of proline, alanine, and glycine residues. Preliminary experiments with alkaline borohydride reduction have shown that about two third of the serine and threonine residues are involved in the protein-carbohydrate linkage, which means that one amino acid over three is involved in this linkage.

The carbohydrate moiety seems to include short chains composed of one galactose residue at the nonreducing end and one N-acetyl-galactosamine residue linked to serine and threonine, and long chains composed of one sialic acid residue at the nonreducing end, one N-acetylgalactosamine residue linked to serine or threonine, and two or three residues of galactose and one of N-acetylglucosamine. The relative number of both types of chains may vary with the conditions of the extraction. The structure of the short carbohydrate chain is of interest because a similar structure was proposed by Dr. G. Uhlenbruck in 1971 for the antigenic determinant of Vicia graminea lectin. In collaboration with Dr. G. Springer, we were able to demonstrate that the glycoprotein of the TA3-Ha subline inhibits the agglutination of NN-blood-group active human erythrocytes by V. graminea lectin to the same degree as the most active preparation of N-antigen. We have used this property for studies that I will discuss later.

Further examination of the high-molecular-weight glycoprotein by Dr. Lodington in collaboration with Dr. H. Slayter showed that it was composed of fragments having a molecular weight between 100,000 and 500,000 daltons, which behave like stiff, straight rods. Since these fragments have been obtained by proteolytic digestion, one can estimate the molecular weight of the original glycoprotein to be in the range of 500,000 to 1,000,000. These physical properties may be important to explain the different biological properties of the two sublines.

In a study just published Drs. Codington and Cooper took advantage of the specific inhibition of the V. graminea lectin by the light-molecular-weight glycoprotein to detect the presence of this compound in the ascites fluid and serum of the Ha-bearing mice, and its absence in the fluids of the St-bearing, syngeneic mice. By use of gel filtration, they could show that the glycoprotein isolated from the ascites fluid had a molecular weight higher than that of the fragment released by trypsin. However, the carbohydrate compositions of both glycoproteins are practically identical;

and we may assume that the molecules released into the ascites fluid are the original molecules embedded in the plasma membrane. In work just completed, we could observe that the material present at the surface of the TA3-Ha cell and reacting with the V.graminea pectin was, in fact, the same material isolated as high-molecular-weight glycoprotein after proteolytic digestion. Other molecules present at the surface of the cell contributed to only ten percent of the reaction between V.graminea lectin and Ha cells. It is of interest that removal of sialic acid residues increased the reaction with the lectin, thus showing the masking effect of adjacent carbohydrate chains. As expected, the strain-specific St cell showed practically no inhibitory activity with V.graminea lectin, the level observed being approximately equal to that of the trypsin-treated Ha cell. Although the exact chemical structure responsible for inhibitory activity against V.graminea lectin is not yet completely known, this activity is a very convenient tool to detect the presence of the high-molecular-weight glycoprotein in Ha cells and in the host bearing these cells and to study the variations of its content in cultured cells.

The presence of the high-molecular-weight glycoproteins at the surface of the Ha cell and its absence at the surface of the St cell may explain the masking of the H-2 antigens at the surface of the Ha cell. Experiments done in our laboratory as well as by Dr. Friberg at the Karolinska Institute have shown that the H-2 antigens can be unmasked by freeze-drying of the cells, which probably causes a collapse of the aggregate of stiff, rod-like long molecules. Unfortunately, it has not been possible until now to visualize these high-molecular-weight glycoproteins in the electron microscope and ascertain their distribution. The high density of carbohydrate chains along the stiff molecules may be more responsible for their distribution at the surface of the cell than components located in or below the plasma membrane.

REISFELD: Are there any comments or questions?

KENNEL: Are either of these cells viral producers?

JEANLOZ: Examination by electron microscopy done by Dr. J. P. Revel has shown numerous viral particles inside the TA3-Ha cells; but we have never observed budding and formation.

KENNEL: What type of virus can you detect?

JEANLOZ: We have not been able to identify a virus. However, we have observed recently in our laboratory a transformation from St to Ha cells similar to that which took place more than twenty years ago at Roswell Park Institute. Unfortunately, the cause of the transformation has not as yet been identified.

KENNEL: Is there any chance that the virus is a source of the glycoprotein that you find on the surface?

JEANLOZ: I don't follow your question.

REISFELD: In other words, what you find on the cell surface, is this really a viral glycoprotein?

JEANLOZ: In general, budding viruses are supposed to take away part of the glycoprotein coat of the host cell. In our case, we do not see budding formation. It should also be noted that we start with a cell already transformed, eventually by virus infection, and what we study is a secondary transformation, eventually caused by another virus.

KENNEL: In the situation of murine leukemia and lymphoma, if one looks at the surface of lymphocyte cultures, established from New Zealand Black animals, one finds that the cultures produce tremendous amounts of a leukemogenic agent. This is a classic murine leukemia virus, and these viruses contain glycoproteins on their surfaces. We have found this glycoprotein by labeling the surface of the lymphocytes in culture with lactoperoxidase. There possibly is enough virus sticking to the surface of these cells that we are labeling and then isolating glycoprotein from the virus. We know for a fact that the glycoprotein is a component of the viral surface. Hans Marquardt in Ralph Reisfeld's lab has purified this glycoprotein from virus. But in addition to that, we have found evidence that this glycoprotein is also expressed on the surface of the cell. Whether it is there due to minute viral contaminations or whether it is there as a component of normal and/or leukemic lymphocytes, it constantly is integrated into the viruses.

KAHAN: I suppose one question that comes up is whether the isolated antigens have a bit of virus, or possibly H2. Is there any possibility that the protein backbone on which the sugar is being added is actually the H2 antigen? Have you tested whether any of your glycoprotein fractions have any H2 activity? In other words, is antigen-masking actually going on at the molecular level by adding these sugars?

JEANLOZ: We have not tested our fractions for H-2 antigenicity.

CUNNINGHAM: How much H-2 antigen do these cells have on their surface as compared to lymphocytes?

JEANLOZ: I remember that Dr. Sanford estimated an amount in the same order of magnitude.

REISFELD: Any more comments or questions? If not, I think we will go on to have Steve Kennel tell us about some of his work.

KENNEL: There are just a couple of comments I would like to add for those who are using lactoperoxidase for labeling cell surfaces. One of the problems we got into is that lactoperoxidase will label anything that is in the final washed cell preparation. As I have just mentioned, concerning the lymphoblast lines we are working with in the mice, the virus apparently sticks to the cell surface long enough to get labeled as "cell surface components," but we really should not consider it to be cell surface. Another entity that is also a problem in cultured cells is fetal calf serum components which actually may stick preferentially to certain cells. For instance, in human lymphoblast lines immunoglobulin G from fetal calf serum is more selectively absorbed from the calf serum in which the cells are grown and can be labeled and isolated as membrane IgG if one does not take care to exclude it. Another caution about the use of the lactoperoxide would be that there have been certain attempts to quantitate amounts of a particular protein which is labeled and subsequently isolated by immunoprecipitation. In all iodination schemes there is apparently a selective iodination of particular tyrosine, and unless one has a purified protein in which one can determine the number of counts per microgram of protein, there is absolutely no way one can make a quantitative statement about the isolated molecules. Finally, in particular circumstances lactoperoxide may label itself or one of its neighbor enzymes, or whatever. The upshot is that in cell preparations that are not adequately washed one has a contaminating 77,000 dalton glycoprotein-lactoperoxidase which is labeled very efficiently. This type of problem has to be controlled. Furthermore, the enzyme isolated from bovine milk is in the oxidized form. When one adds the enzyme to a cell preparation and then adds radioactive iodine, it is at least theoretically possible to get a complete round of labeling without any exogenous oxidizing agents such as peroxide.

      Dr. R. Levy and I in Dr. R.A. Lerner's laboratory have encountered this problem. After we had finished the labeling reaction, we added exogenous lactoperoxidase and then, in performing immunoprecipitation of lactoperoxidase, we found out that the lactoperoxidase in fact labeled the antibody that we added subsequently. The enzymes, when added in the oxidized form, picked up some radioactive iodide and put this on the antibody due to its close proximity. I would like to state that this occurs under peculiar circumstances, but some experiments are designed in that way and we really need to be cognizant of particular pitfalls. This enzyme catalyzed radilabeling technique is indeed very, very sensitive.

REISFELD: I think that these words of caution should provoke some discussion, since this particular method of cell surface labeling is most popular among many immunologists today.

DAVID: Actually, I have several comments. First, regarding your concern about labeling cell-bound fetal calf serum components, can you possibly displace them by washing the iodinated cells with cold fetal calf serum?

KENNEL: We have also controlled for those that are present.

DAVID: I was just wondering how tightly these materials are really bound. I have done some cell labeling and I usually include a wash with fetal calf serum at the end but don't really know for a fact whether this helps.

KENNEL: For some reason or other, if one labels the surface of the human lymphocytes, one finds the labeled fetal calf serum in the first wash of those cells even though the lymphocytes have previously been washed exhaustively. There is something about the oxidation and the labeling which ehlps to release these proteins. During subsequent incubation at 37° one can lose some more fetal calf serum proteins.

DAVID: How do you know these are fetal calf serum components? In other words, during incubation at 37°C at lot of cell surface components can be lost, can they not?

KENNEL: We look at them as antibody precipitates.

DAVID: This is more a comment than a question. I think I heard you suggest that labeling you saw at one point was due to lactoperoxidase being isolated in its oxidized form. In fact, I think certain people have suggested that the oxidized enzyme is not stable and might have a tendency to self-destruct. As it turns out, I can get much more than a single round of iodination without exogenously added peroxide. One usually carries out the iodination reaction in the vicinity of 10^{-8}M enzyme. We can iodinate IgG quite efficiently without adding hydrogen peroxide if we use iodide concentrations around 10^{-6}M. It might be that normal buffers contain relatively high peroxide levels, or that oxidase activity is present, but the point is that you can get considerable enzymatic iodination without having to add peroxide at all.

KENNEL: I am not particularly married to the explanation. I simply wanted to caution that once one is finished with labeling what one wants to label, it is important to add something to inhibit the enzyme, e.g., merthiolate, or azide, or whatever fits the particular system.

SUNDHARADAS: I have done some iodination experiments with intact lymphocytes and I find only 20 to 30 percent of the counts incorporated into proteins, and I have the feeling that quite a bit of the rest goes into the lipids, probably as dissolved free iodine. Is it possible that the free iodine can non-enzymatically iodinate the proteins?

REISFELD: This is certainly an interesting technical point and important in a way because many conclusions that one makes are based on these techniques. It is obviously important to be very careful, as it was pointed out in this discussion.

UHR: I think that all the points Steve Kennel raised are pertinent ones, and I also agree with the last comment that lipids can be labeled with enzymatic iodination. However, the implication of the chairman's comments are that these cautionary notes represent major problems with the method. Of course, the method has restrictions and pitfalls. It is the method, however, for incorporating extraordinarily high levels of radioactivity into a handful of molecules on the cell surface thereby facilitating biochemical analyses of these molecules. Further, I don't see that the problems raised by Kennel are very serious ones. Labeled lactoperoxidase is simply washed away. The murine cells we have looked at have not displayed detectable binding of cytophilic proteins. In fact, I have been concerned at the virtual absence of IgG which can be labeled by enzymatical iodination on splenocytes. We apparently wash off the cytophilic immunoglobulin or release it by the iodination reaction. I would not want the non-immunologists left with the idea that this is a technique of very limited usefulness. I think the method has great potential.

POLLARD: We* have made an effort to study fusion dynamics of secretory cell membranes by isolating a complex of chromaffin granules and plasma membranes from the bovine adrenal medulla.

I. Membrane structure and membrane fusion.   Many current membrane models stress the fluid-like character of membranes, and relate this quality to a common phospholipid bilayer. There are many intuitive bases for this idea. Membranes are believed to be able to fuse, or "flow together" like molten metals. Examples of the fusion process are the mending of broken membranes (for example, osmotically intact red cell ghosts), or exocytosis, where a hormone or neurotransmitter is released by the fusion of a secretory vesicle with the plasma membrane. Membrane fusion during exocytosis has been observed by electron microscopy in at least 18 different cell systems (Poste & Allison, Bioch. Biophys. Acta (300:421-465, 1973).

_____

* This work was done in collaboration with Drs. O. Zinder and P. G. Hoffman.

The physical properties of many of the most conveniently obtained plasma membranes are consistent with the existence of a fluid phospholipid bilayer, and virtually no other long range organization. Membrane proteins appear to diffuse much faster than if the medium they were associated with were a solid (Frye and Edidin, J. Cel Sci. 7:319-335,1970; Cone, Nature New Biology 236:39-43, 1972; Hubbell, et al., Proc. Nat. Acad. Sci. 64:20-26, 1969). Electron microscopy and X-Ray of red cell membranes and myelin have been consistent with this hypothesis (Branton, Ann.Rev.Plant.Physiol. 20:209-268, 1969; Wilkins, Ann.N.Y.Acad.Sci. 195:356-365, 1972).

From this perspective, fusion ought to be a commonly observed property of most membranes. But quite the opposite is true. There exist only a few rudimentary systems appropriate for the study of membrane fusion at the molecular level. The fusion of Sendai virus with plasma membranes of cells has received attention (Apostolov and Poste, Microbios.6: 247-261, 1972), as has the possible fusion of liposomes (Rasmussen, Science 170:404-412, 1970). No exocytosis system has been shown to work in vitro, to our knowledge. The most recent work on this problem suggests that the fusion process in real, biological systems is energy dependent and probably requires additional chemical signals for membrane recognition and specificity (Poste and Allison, Bioch. Biophys. Acta 300:421-465, 1973).

It is becoming obvious that the quite general membrane models featuring a "sea of fat" with "protein boats" upon it have little predictive value for highly specific events like fusion. In addition, membranes containing highly ordered protein units, in addition to phospholipid bilayers, have been discovered. These include the purple membrane (Blaurock and Stoeckenius, Nature 233:152-154, 1971), the gap junction (Goodenough and Stoeckenius, J. Cell.Biol. 54:646-656, 1972), and the chromaffin granule membrane (Pollard, et al., J. Supramol. Struct. 1(4):295-306, 1973). It seems evident that a new generation of membrane models may be forthcoming when better information is known about specific membranes and membrane functions. This, of course, would be analogous to the revolution in our understanding of proteins following the recent progress in X-Ray diffraction.

II. Exocytosis and Membrane Fusion in the Adrenal Medulla. Most of our fundamental information about the process of exocytosis is based on studies on the secretion of epinephrine from the adrenal medulla. The chromaffin granule (or secretory vesicle) is known in some biochemical (Banks, Biochem. J. 95:490-496, 1965; Pollard, Fed. Proc. 32(3): 525, 1973) and physical (Pollard, et al., J. Supramol. Struct. 1(4):295-306, 1973; Smith, et al., Science 179:79-82, 1973; Plattner, et al., Ultrastructure Res. 28:191-201, 1969) detail. The importance of calcium for secretion and the concept of "stimulus-secretion coupling" were first recognized in this system. Since large amounts of homogeneous tissue could be readily obtained, it seemed that this might be a favorable system for the study of the chemical and molecular basis of membrane fusion. We, accordingly, focused our attention on the plasma membrane of the chromaffin cell.

1. Electron Microscopy of Chromaffin Cells. Chromaffin cells of most vertebrates are filled with dense core vesicles (chromaffin granules), as shown in Figure 14, A.

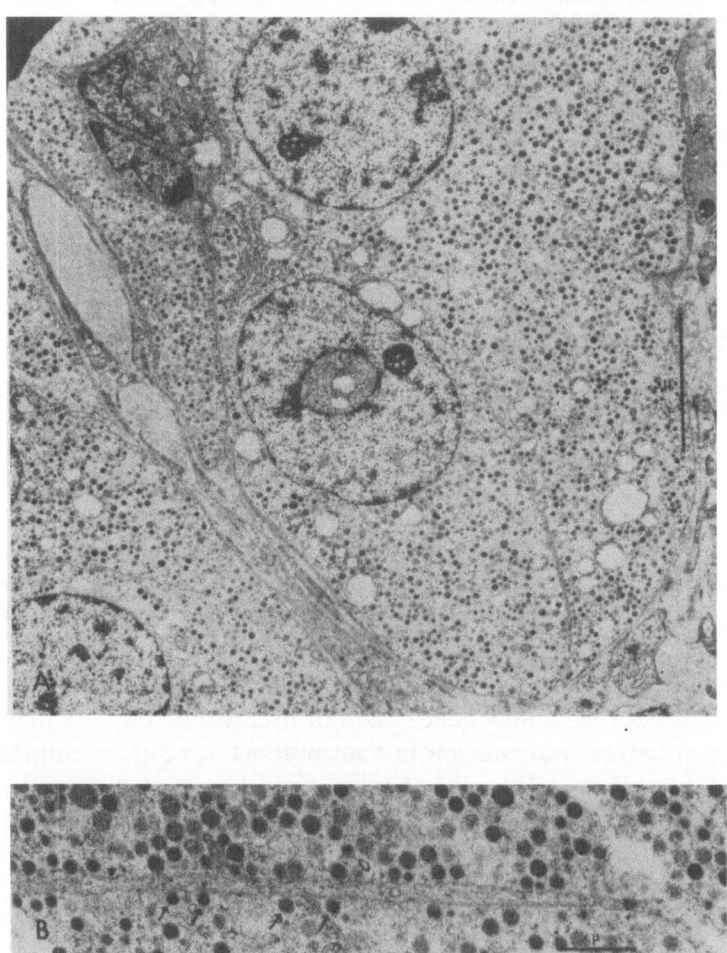

Fig. 15. Electron Micrographs of Adrenal Medulla from Guinea Pig. Animals were decapitated and the tissue fixed in 2.5% glutaraldehyde, post-fixed in 1% osmium textroxide and stained <u>en bloc</u> with 1% aqueous uranyl acetate. A. Low power view (7200X). Cells are filled with dense core vesicles (chromaffin granules). The granules show variability with respect to density of the core. B. Higher power (21,000X).

Chromaffin granules a r e shown to be closely juxtaposed to the plasma membranes of the chromaffin cells (see arrows). Examination of the plasma membranes of these cells revealed a large number of chromaffin granules (CG) in close juxtaposition (Figure 15, B, arrows). Since these two membranes presumably fuse during exocytosis, it seemed that a study of the isolated plasma membrane (PMb) might yield useful information on the fusion process.

2. Isolation of the Plasma Membrane. Isolation of a plasma membrane-rich fraction proved to be relatively simple, using differential and equilibrium gradient techniques. The m a r k e r enzymes for plasma membranes (5''-nucleotidase, Na^+, K^+ATPase, adenylate cyclase) we r e found to be enriched in the region between density 1.08 and 1.14, sucrose (Table 2):

TABLE 2

ENZYMATIC IDENTIFICATION OF MEMBRANES IN BOVINE ADRENAL MEDULLA

Enzyme (units)	5'-nucleotidase (u moles/mg/min)	Na^+,K^+ - ATPase (n moles/mg/min)	Adenylate Cyclase (p moles/mg/min)	Monoamine Oxidase (u moles/mg/min)	β-glucuronidase (u moles/mg/hr)	Dopamine-β-hydroxylase (p moles/mg/min)	Mg^{++}-ATPase (n moles/mg/min)
I 1.04/1.08	1.12	18.2	17.0	N.D.	3.00	47.0	29.6
II 1.08/1.14	1.58*	36.7*	31.3*	1.04	5.84	155.5**	133.9*
III 1.14/1.21	0.95	17.7	18.8	11.76*	12.56*	49.5	47.7
IV 1.21/1.29	0.10	7.6	15.3	1.72	7.60	1100.0*	41.1

Mitochondria a n d lysosomes were significantly reduced in this region. However, it wa s noted that dopamine-B-hydroxylase (DBH), an enzyme marker for the chromaffin granule membrane, was present in significant amounts. In separate experiments, it was found that epinephrine was also present. Well over 90% of the remainder of the chromaffin granules (DBH, epinephrine) were located in their customary high density region.

3. Protein components of the Plasma Membrane. The protein components of the plasma membrane (PMb) were examined by discontinuous SDS gel electrophoresis, and c o m p a r e d with p u r i f i ed chromaffin granule membranes (CGMb) and granule lysate (see Figure 15).

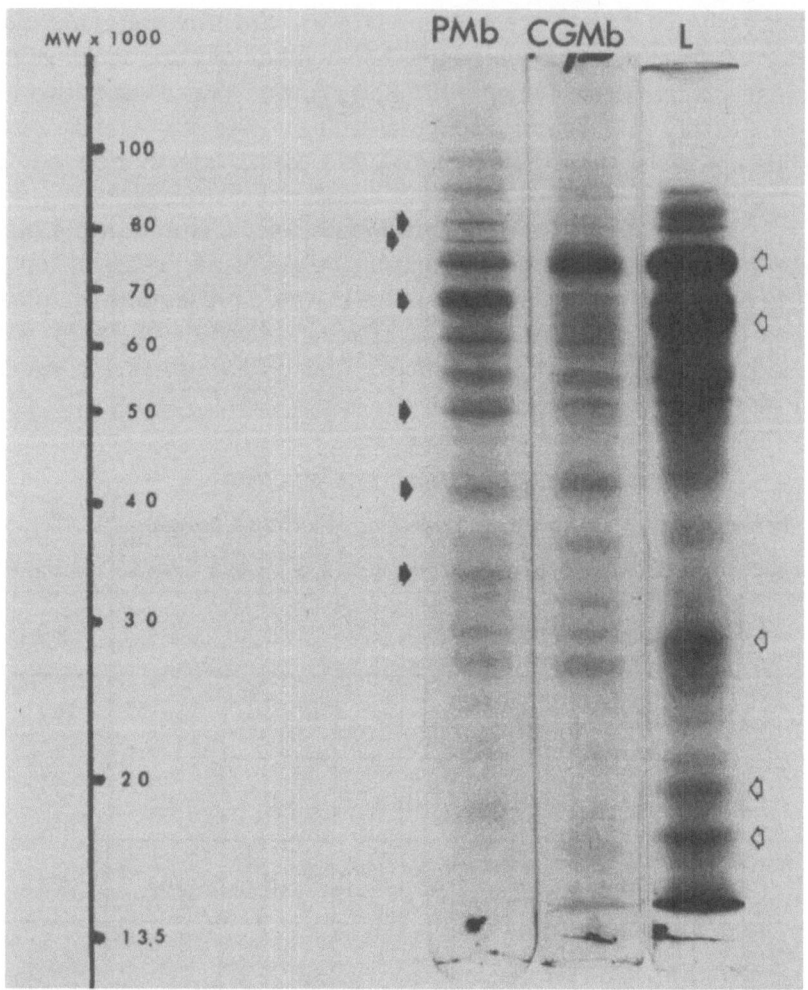

Fig. 16. Discontinuous SDS gels of Bovine Plasma Membranes (PMb), Purified Chromaffin Granule Membranes (CGMb) and Soluble Proteins from Chromaffin Granule Lysate (L).

Gels were run according to Laemli with 15% T and length=12cm 30 μg protein were applied to each sample. Both the PMb and the lysate (L) fractions were compared to the CGMb. The PMb had six clearly unique bands (solid arrows). The chromaffin granule lysate (L) had five unique bands (open arrows), including the major species, chromogranin A (77,000 MW).

All bands found in the CGMb were also found in the PMb. However, as indicated by the solid arrows, at least six clearly unique bands could be identified in the PMb that were absent from the GCMb.

The PMb fraction was also analyzed for cytochrome b_{562}, a specific membrane component of the chromaffin granule. Utilizing both the alpha and gamma bands of the reduced-minus-oxidized difference spectrum at 22° C, it was concluded that the PMb was at least 50% CGMb.

4. Electron Micrographs of Isolated Plasma Membranes. An electron micrograph of the isolated plasma membrane fraction is shown in Figure 17.

The fraction proved to be highly heterogeneous, but three major classes of structure could be discerned. Large vesicular membranes (L) of approximately one micron in diameter were frequently found. These large vesicular membranes were often found to contain chromaffin granules. By analogy with synaptosomes from nervous tissue, we called the latter structures "medullasomes."

Finally, many examples could be found of intact chromaffin granules attached to larger membrane fragments. The attachment point was often quite electron dense (see arrows). These structures were similar to the close juxtaposition of chromaffin granules and plasma mem- observed in the intact tissue (Figure 15 B).

These data suggested that a complex between chromaffin granules and plasma membranes existed and could be isolated physically. A new and perhaps more general question could now be asked: what held them together? What was the basis for the specificity?

5. Glycoprotein Components of the Plasma Membrane. Glycoproteins are known to be responsible for recognition processes in a variety of biological systems, and in addition, are believed to penetrate biological membranes. We decided therefore to examine the plasma membrane glycoproteins and compare them with those of the chromaffin granule membrane. The glycoproteins were detected by staining discontinuous SDS gels of the two membranes for carbohydrate, as described in Figure 18.

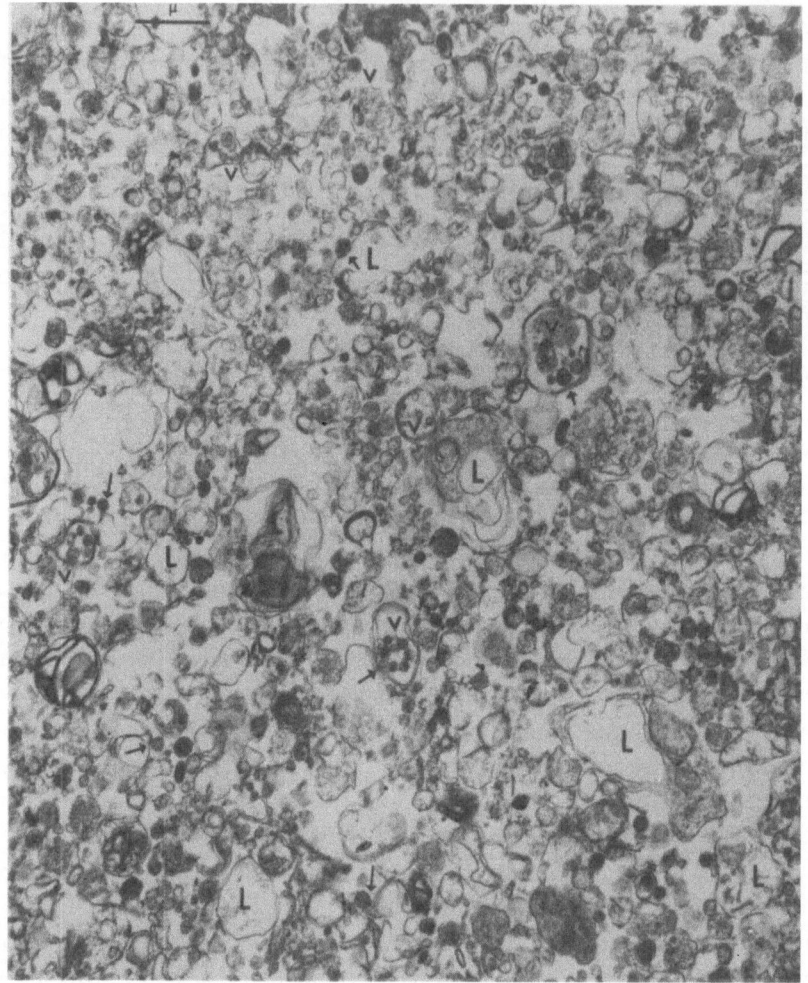

Figure 17: Electron Micrograph of the Isolated Plasma Membrane Fraction of Bovine Adrenal Medulla (17, 500 X).

Three classes of structure are observed: Large, closed vesicular sheets (L) of about one micron in diameter and closed vesicular sheets containing dense core vesicles (V) are found. Dense-core-vesicles closely attached to the membrane fragments (arrows) are observed. The juxtaposition of dense-core-vesicles and plasma membrane is similar to that observed in the intact cells (Figure 15B).

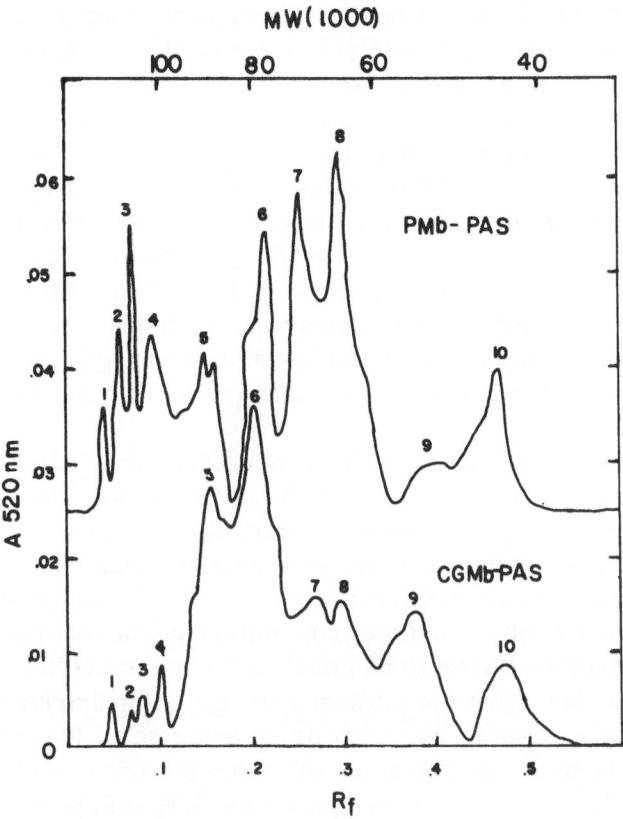

Fig. 18. Desitometric Scan of PAS-Positive Glycoproteins from Plasma Membranes (PMb) and Chromaffin Granule Membranes (CGMb) of the Bovine Adrenal Medulla.

Samples of membrane (300 μg protein of CGMb and 600 μg protein of PMb) were electrophoresed in discontinuous SDS gel according to Laemmli and were stained for PAS positive material as described by Marchesi. Gels were 10 cm long, and were scanned in a Gilford Gel Scanner at 520 nm. Omission of the periodate step resulted in loss of all bands. The PMb trace is displaced upward by an arbitrary amount to permit detailed comparison with the CGMb trace. Peaks in both fractions, thought to be identical, are given arabic numerals. The gels of PMb and CGMb, when stained by PAS, appeared identical when examined by eye.

When examined visually, the two gels appeared to have identical band patterns. Densitometric analysis verified that all ten peaks identified in the chromaffin granule membrane fraction. In qualitative terms it took twice as much plasma membrane fraction as pure chromaffin granule membranes to give equivalently stained P.A.S. gels. These data appeared to coincide with the earlier quantitative measurement of the cytochrome b_{562} content of the plasma membranes (where it was calculated that the plasma membrane was 50% enriched in chromaffin granule membranes).

These data tended to suggest that all of the detectable glycoproteins present in the plasma membranes were of chromaffin granule origin. Such information lends support to the hypothesis that glycoproteins might be involved in recognition processes leading to membrane fusion.

III. Plasma Membranes from Adrenal Cortical Cells: A Test for the Uniqueness of the Adrenal Medulla Data. In the foregoing section chemical evidence was presented in support of the contention that the plasma membrane of the chromaffin cell formed a complex with chromaffin granules, and that glycoproteins might be involved in membrane recognition. However, the data would be necessarily ambiguous unless the characteristic patterns could be shown to be specific for the medulla. Therefore comparable fractions from the adrenal cortex, a neighboring and potentially contaminating tissue, were similarly examined. Plasma membranes were prepared from cortex tissue by the same method used for medullary cells and examined by Laemmli discontinuous SDS slab gels. The cortical plasma membrane had 30 discernible bands of which six were clearly absent from the medulla plasma membranes. The medulla plasma membranes had 31 discernible bands of which seven were clearly absent from the cortex. The soluble proteins (supernatant from a 200,000 X g X 3 hours centrifugation) differed to a similar degree (about 20 %). Only about 50% of the chromaffin granule membrane proteins overlapped with proteins in the cortical plasma membranes. Nonetheless, it was somewhat disconcerting to discover the significant degree of corresponding bands in the several membranes. In order to assess the potential contamination of medulla in cortex and vice versa, epinephrine and ACTH-activated adenylate cyclase were measured in each fraction. Whole adrenal cortex had only 3% of the epinephrine levels that whole medulla contained, and crude plasma membranes from the cortex had no detectable epinephrine. The adenylate cyclase from the adrenal cortex plasma membrane was activated approximately 8-fold by 10^{-4}M ACTH, while the medullary adenylcyclase was essentially unaffected. These data convinced us that the cross contamination between tissues was essentially nil, and that the membranes from otherwise diverse origins shared many protein size classes.

It was also crucial to examine glycoproteins from the adrenal cortex. The glycoproteins from the cortex were found to be fewer in

number (only four) and to differ in mobility in every case when compared
to medullary membrane glycoproteins. The data presented here demonstrate that chromaffin granules from the bovine adrenal medulla are capable
of forming complexes with plasma membranes from the same cell. Perhaps because of the formation of the complex, some chromaffin granules
appear to be trapped within plasma membrane vesicles. These we have
termed "medullasomes," by analogy with synaptosomes. The chromaffin
granules forming complexes with the plasma membranes appear to be intact. The evidence for this is that they are, in general, dense-core
vesicles, and that the fraction contains epinephrine. This suggests that the
process of membrane fusion involves a preliminary step of membrane contact that appears to be distinct from the actual union of membranes. The
present data is consistent with a hypothesis that the specificity of this contact may involve glycoproteins of the chromaffin granule membrane. However, other mechanisms must be considered: calcium may mediate the
union; fibrous proteins such as actin, myosin, or tubulin may be involved.
We are actively pursuing these questions.

The fundamental importance of the work presented in this paper
rests in the possibility that processes involved in membrane fusion during
exocytosis can now be studied at the molecular level in a homogeneous system. A number of important chemical questions relating to membrane
recognition and fusion can now be asked.

EDELMAN: Have you tried to reconstitute the complex, and have you tried
to see if they are reconstituted only on one side of the membrane?

POLLARD: We are interested in reconstitution experiments, obviously.
The first question is, can we take apart the plasma membrane vesicle
complex. We tried to dissociate the complex by taking the calcium away.
But this was not successful. We lowered the pH to 3, and that did not do
it. Finally, we raised the ionic strength to 3.36 M KCl, and had some
success. We are now pursuing the question about whether or not the plasma
membrane will in fact interact with secretory vesicles, and we have some
suggestive evidence that perhaps it does. But it is all very preliminary at
this point. We also want to ask other questions now such as whether or not
molecules like colchicine, various local anesthetics, or the chlorpromazines
can affect the formation or dissociation of the complex. I think this system
will be a very useful system for looking at biologically relevant interactions
of membranes. I want to point out one particular property of the chromaffin granule membrane. It is very highly ordered membrane, and quite
different from the usual plasma membrane system. It is highly ordered
in terms of protein subunits, as defined by freeze fraction and X-ray.
When the chromaffin granule and plasma membranes fuse, the plasma
membrane must be ready to take something highly ordered. This is

something we can relate very directly to what Dan Branton was talking about today with regard to ordered structures underlying apparently disordered submembrane particle distributions.

BRANTON: You imply in your argument that glycoproteins act as recognition sites for vesicle contact and the glycoprotein would be on the inside surface of the cell membrane and on the outside surface of the chromaffin granule. That is unusual, if it is true.

POLLARD: My understanding is that the literature predicts that glycoproteins on plasma membranes stick out, right?

BRANTON: Yes

POLLARD: Let me draw a chromaffin granule membrane and let me draw a plasma membrane over here. Let me represent the sugar as a christmas tree sticking out. Supposing you have a similar surface on the chromaffin granule membrane with christmas trees sticking out as well. I can visualize that the sugar on the chromaffin granule membrane might find a receptor on the plasma membrane, and then go right on through. That is one idea, it's just a fantasy, okay? The second thing we know is that there is at least one glycoprotein inside the vesicle, and that is DBH. So a second fantasy you could consider is the possibility that these glycoproteins may be involved in bringing the recognition system up close, but when the actual fusion occurs, that what is inside is now going to be outside.

BRANTON: Right.

POLLARD: So internal glycoproteins might be readily apparent in this system. In fact, we have tried to label glycoproteins in intact vesicles and in the broken membranes. In broken membranes you can label all of the glycoproteins where in enclosed vesicles you can label only some.

BRANTON: That is the usual situation. What I am asking about, is this postulated glycoprotein interaction? The sugars act in recognition here and I don't know of any case in which glycoproteins have been found on the external surfaces of internal or the cytoplasmic surfaces of vesicles inside the cell.

POLLARD: That's the thing, very few plasma membrane-like glycoproteins are found in internal membranes. These particular glycoproteins on the chromaffin granule membranes can be easily isolated by LIS extraction, except for the DBH. That is the only evidence for a plasma-membrane like quality for these membrane glycoproteins. It may be a crummy criterion.

But a curiosity remains to be discussed. It could be that those secretory vesicles are precursors of plasma membranes, or something like that. That is the notion we are pursuing. We have to have some way for this vesicle to recognize that there is a plasma membrane. This is one way.

REISFELD: Even before LIS was mentioned, Marchesi wanted to make a comment.

MARCHESI: I must tell you that I am very skeptical of some of your findings. Have you ever done the following: have you mixed isolated chromaffin granules with any other membrane system and homogenized it to make sure that this is not an artifact? You know as well as I do that it is very easy to produce material incarcerated inside vesicles?

POLLARD: We did not do it with cortical membrane.

MARCHESI: You realize that this is a very difficult experiment to do because you can never recreate the proper conditions. These things which may open and close may already have closed by the time you put the exogeneous granule in. I am not saying it is easy, but it is a problem you should consider. Secondly, you have to be careful about the use of PAS stains. You are showing essentially all the polypeptides as being PAS positive.

POLLARD: No, these stains represent what happens when you leave out periodate as a control.

MARCHESI: Many people in the past have discovered that PAS will bind to things that are not glycoproteins. Also, we know that there are proteins in the red cell that have maybe five or 10 percent sugar which do not stain at all. Finally, I wouldn't make any generalizations about these membranes, they are too crude and too many things could be in them.

POLLARD: That was the main conclusion I had, based on the fact that you could not easily distinguish one plasma membrane from the next with SDS gel except on the basis of a small number of bands, 20 percent. This is too small.

MARCHESI: I think we should also state that there is still a controversy as to whether or not there is any sugar on the inner surface--just because it does not show up by the usual procedure does not mean it is not there.

POLLARD: No.

HAKAMORI: I feel that I must say something about some of the glycolipids and glycoproteins that could be exposed inside of the membrane (cytoplasmic side), because of our experiments which indicate that there is an increased quantity of glycolipid and glycoprotein labile in ghosts as compared to intact cells.

RAFF: There is something very rabbit-out-of-the-hat about your model for chronaffin granule recognition of the plasma membrane.

POLLARD: I made it up.

RAFF: What is the evidence for glycoproteins on the surface of the granule membranes and on the cytoplasmic side of the plasma membrane?

POLLARD: Which ones? If you are referring to the outside of the chromaffin granule membrane, there is no evidence. But, in addition, there is no evidence that there is a glycoprotein facing the outside of the plasma membrane. I just assumed that might be the case, based on studies on the red cell.

RAFF: Are you suggesting the glycoprotein is on the cytoplasmic side of the plasma membrane?

POLLARD: Yes, the outside. The next thing is, this is the cytoplasm.

RAFF: Do you want to have glycoproteins interacting with each other on the cytoplasmic side?

POLLARD: No.

RAFF: Then how do you see the recognition taking place?

POLLARD: How do you get recognition between con-A and α-methyl mannoside? I assume there must be some protein or receptor for the vesicle for the glycoproteins.

RAFF: What is the evidence that the sugar exists there?

POLLARD: The evidence is twofold. First of all, you can identify glycoproteins found in the plasma membrane fraction as being identical to those of the chromaffin granule membrane. You can also interpret the data to show that the plasma membrane glycoproteins are probably from the chromaffin granule membrane. Secondly, you can extract those glycoproteins into Lis. You can also label the glycoproteins with iodine and

and extract the counts into Lis. Since this experiment also works on intact chromaffin granules, it is possible that some of those proteins may be exposed to the exterior. I am not really saying that the glycoprotein is specifically on the outside. I have only given some reason for considering it as a hypothesis. I have not given any evidence, compelling to myself, that that is exactly the way it is. It is a possible interpretation which I have placed on the data.

RAFF: The second question I wanted to add is this: Isn't the magic of these secretion systems that the secretory granules don't in fact fuse with plasma membrane until calcium enters the cycle and triggers the whole thing off?

POLLARD: Yes.

RAFF: So why would you want to have this proposed recognition system operating before the cell is activated?

POLLARD: I think I tried to point out when you look at electron micrographs of the intact tissue and if you look at the plasma membrane, you see secretory vesicles juxtaposed to the membrane anyway. They are just sitting there. Perhaps the association is a bit like ribbon synapses where vesicles are sitting very close to the plasma membrane, just ready to move when stimulated. In other words, just looking at tissue, you see these things lined up on plasma membranes. The data I showed you are not unique. Other people have found similar pictures, so the condition of actually juxtaposition of the secretory vesicle and the plasma membrane is not something I made by homogenization. It is something which I have found in my homogenate, which I have also found by just looking at the tissue.

RAFF: But the majority of granules are not associated with the plasma membranes.

POLLARD: That is right, well over 90 percent are not associated.

LERNER: I would have thought just the opposite. Why would you assume that the plasma membrane of different somatic cells would be very different? After all, in the last analysis the idea is to keep the inside from becoming the outside. During evolution, a good structure evolves and it remains basically the same. Keep in mind that most differentiated phenotypes involve a very small percentage of gene expression of the cell. One can superimpose a great deal of differentiation onto a common basis, so thus I would have predicted you would have obtained just the result that you did. In short, I would say that most somatic cells are more alike than different.

POLLARD: I thought when we started doing this that all the protein components of plasma membranes from different cells in the same animal might be very, very similar. I mean, I believe in the fact that the membranes came from one original cell and they were modified to suit the various purposes of different tissues.

LERNER: That is what you found?

POLLARD: Right, but the interesting thing is that the secretory vesicle membranes are not different; in fact, they are very similar in terms of their proteins to proteins of the plasma membrane. Perhaps they were manufactured in the same place.

STECK: I don't see the necessity of generalizing on the meager information we now have on the topography of glycoproteins in a few systems to all cells at this point and saying that because we don't find sugar on the cytoplasmic surface of this or that cell, the red blood cell, for example, that it cannot be true in the adrenal medulla. For example, the rhodopsin of retinal rod disks, it is claimed, has its oligosaccharide moiety on the outside of the disk, which is opposite in orientation to that presumed from studies on the endoplasmic reticulum or Golgi apparatus. However, I do not believe that Harvey has sufficient data on his membranes to pinpoint orientation. With respect to Dr. Hakamori's comments, we have looked at the two surfaces of the human red blood cell membrane for sugar contained in glycoproteins and glycolipids using galactose oxidase plus tritiated borohydride. Our evidence suggests that <u>all</u> of the labeled species are localized to the external and not the cytoplasmic surface.

HAKAMORI: Galactose oxidose treatment followed by reduction with tritiated borohydride is a method which is not relevant to depict how much is exposed or not exposed or what percentage of the glycolipids is labeled or not. More exactly, we should determine how much of the galactose present in glycolipid is destroyed by surface membranes. By applying a technique which utilizes the hydrazide derivative we can see what proportion is exposed or not.

REISFELD: Do you want to attempt to clarify this inside-outside controversy?

POLLARD: Yes, I do. I was making two classes of hypotheses: one relatively conservative, the other obviously wild. The conservative hypothesis is based on the fact that the chromaffin granule and plasma membrane can touch without fusion actually occuring. These data suggest that there are at least two stages in membrane fusion. One involves recognition; the second, and perhaps energy-dependent step, is fusion <u>per se</u>. The second set of interpretations having to do with the glycoproteins is obviously wild, and that is obviously a hypothesis for testing. The only compelling evidence that this could be true, I believe, would be direct labeling of sugars by some means that you could actually see, using an electron microscope and finding they were on the inside or outside of these vesicles. That is what I would consider compelling evidence of that hypothesis. But you would have first have to have a compelling reason to do that experiment, and that is the reason that I have set forth the hypothesis.

REISFELD: I think there are more comments, but I would ask you indulgence. We should have some coffee after all, and then afterwards continue this discussion. We will talk about other substances on the surface, histocompatibility antigens, and, believe it or not, some of these are not glycoproteins. We will recess for coffee now.

REISFELD: I realize that there are many questions which remain unanswered from the discussion that went on just before the coffee break. Is Harvey Pollard back or is he still discussing? Jackie Reynolds has a ten-second question that she wants an answer to.

REYNOLDS: It only requires a yes or no. Did I misunderstand you? Did you say that you found traces of something you thought was like spectrin in chromaffin granules? I thought you said that.

POLLARD: I was speaking functionally and not substantively.

REYNOLDS: If you have any function for spectrin, tell us.

(Laughter.)

POLLARD: There is a protein called chornogranin A, which perhaps has a structural role related to the membrane.

REYNOLDS: That is not spectrin.

POLLARD: Under certain conditions we can make it form beautiful fibers.

REYNOLDS: Were you not claiming that you had a substance with 200,000 molecular weight....

POLLARD: Absolutely not.

REISFELD: That should settle this point. While I found the previous discussion both fascinating and controversial, I would like now to get back to
discuss some histocompatibility antigen molecules on the surface of the
plasma membrane. They may be free-floating, anchored, or in a lipid
bilayer. We know that they have considerable biologic relevance. We
shall now hear from Jack Pincus about attempts to solubilize such materials,
and what constitute some of the critical problems in these studies.

PINCUS: In discussing the problems of the solubilization of membrane components in a form in which they retain their biological activity, I will talk
about three membrane activities associated with immunological phenomena
and the problems associated with obtaining each free of the cell membrane.
When working with these types of membrane-associated activities it must
be realized that the fact that one finds an activity in solution that can be
measured, such as the detection of histocompatibility antigens by antibody,
this does not mean that the biological properties of the soluble material
will be the same as those of the membrane-associated antigens. It may
therefore be necessary, in addition to solubilization, to reconstitute with
lipids. Although this has not been given very much attention in investiga-
tions on mammalian cells, studies on bacterial cell membranes, such as
that on the phosphosugar transport system and salmonella O antigen syn-
thesizing system, have demonstrated its importance. We have chosen
to work for the most part with the cultured cell lines since using whole
tissue presents several problems. The fact that an individual organ contains multiple cell types, a heterogeneous membrane population results.
In addition, many membrane preparations from whole tissue are often
contaminated with elements of connective tissue. By using a cultured cell,
such as cultured lymphoblasts, one essentially has a single cell type and
can work with a single membrane.
            The first membrane associated activity I would like to discuss
which we have succeeded in solubilizing is the lectin receptor from a mouse
leukemic cell line L1212. This work was done in collaboration with Dr.
Benjamin Hourani and Mrs. Nina Chace   (Biochem. Biophys. Acta 328:
520 (1973). The aim of this study was to prepare membranes in high
yields and solubilize all of the glycoproteins from these membranes. The
first thing that came to our attention was that there was variability in the
yields of glycoprotein when cells were grown to random densities. Thus,
the agglutinability of cells, grown to different densities, by concanavalin
A was studied. Agglutinability of cells grown to a density of 5.4×10^5/ml
was maximal, decreasing as the cells reached higher densities. At very
high densities the cells are not agglutinable. As agglutination itself is an
active process, these results may not reflect the amount of glycoprotein
on the surface. However, we did find that when we grew the cells to high
densities our yields, by the methods I will describe were very low, but if

but if we grew cells to low densities where they were most agglutinable, we obtained higher and more reproducible yields of glycoproteins. Thus, the criterion of agglutinability could be used to determine the best cell density for obtaining reproducible yields of glycoproteins.

The next problem was making membranes from the cells grown to the proper density. There are a number of methods that have been used, many involving sucrose density gradient centrifugation, a very time-consuming process that usually results in relatively poor yields of membranes. More recently, Burnette and Till have developed a procedure using a two-phase polyethylene glycol/dextran system which permits a rapid preparation of membranes, but which has to be modified for each particular cell. We adapted the procedure for L1210 cells by stabilizing the membranes with magnesium instead of zinc and altering the time and centrifugation of the polyethylene glycol/dextran separation. We could obtain a plasma membrane fraction from $50x10^9$ cells in a period of two and a half hours. Membrane preparations were examined under the electron microscope after staining with 2% phosphotungstic acid. What was usually seen were sheets of membranes and small rolled-up fragments. Occasionally we found vesicles which represents stabilization of membranes by magnesium. We have performed biochemical analyses on these isolated membranes and found 9.3 percent of the total homogenate protein in the membrane fraction, 72% of the sodium potassium activated AT Pase and 4.2 percent of NADH diaphorase, the latter indicating a very low contamination of the membrane with endoplasmic reticulum.

Having achieved this fractionation, we isolated the glycoproteins according to the lithium diodo-salicylate procedure of Marchesi and Andrews which was used for solubilizing the major red cell glycoprotein. The initial product we obtained was contaminated with nucleic acid which, as suggested to us by Dr. Marchesi, was removed with staphylococcal nuclease. The yield of glycoprotein was between 4 - 7 mg/ $50x10^4$ cells. The isolated membrane glycoproteins were examined by SDS polyacrylamide gel electrophoresis. Whole L-1210 membranes were, as expected, quite complex. The isolated glycoproteins exhibited six components, four of which contained carbohydrate and had approximate MW's of 84,000, 63,000, 44,000, 33,000. We also found an additional two components that did not stain with PAS which we have been unable to separate from the other four components. These could either be glycoproteins with a low percentage of sialic acid that does not stain well with PAS, or acidic proteins that are insoluble in phenol or proteins that run as aggregates with glycoprotein.

We performed qualitative sugar analyses on the preparations. They have the typical membrane sugars that Dr. Ginsburg has described. It is of significance that the preparation did not contain any ribose or deoxyribose, indicating that it is not contaminated with nucleic acid. A further absence of glucose indicates no contamination with glycolipids.

We looked at a biological activity of the glycoprotein mixture and it's
ability to inhibit the agglutination of L1210 cells. We tested the inhibitory
ability against three different lectins, Concavalin A, Lens/Culinaris
hemagglutinin, and wheat germ agglutinin. At approximately equal concentrations of lectin and glycoprotein, we observed complete inhibition of
agglutination of L1210 by all three lectins. Thus, at least this carbohydrate associated activity is retained upon isolation. To demonstrate that
the glycoproteins were derived from components on the surface, we took
the isolated material, injected it into a rabbit and made an antiserum which
we then reacted with intact L1210 cells. The antibody was cytotoxic while
the preimmune serum was not, indicating that what we have isolated is
relevant to what is on the surface. This approach is relatively straightforward and is probably useful for most membrane glycoproteins. A more
difficult activity to solubilize has been the membrane proteins associated
with histocompatibility activity. We have not had any success with these
using either lithium diodosalicylate or detergents. A very popular procedure that has been used is 3M KCl, a method developed by Reisfeld and
collaborators. However, it has some peculiarities. The yields of such
soluble antigens from isolated membranes are much lower than when one
uses whole cells. There also seems to be a cytoplasmic factor involved
in the solubilization. Furthermore, there is an eight-hour long lag period
before one begins to see an appreciable amount of soluble antigen.

If one thus performs a hypertonic salt extraction on whole
cells a viscous mass, due to the liberation of nucleic acids, occurs.
This causes problems in attempting further purification and often reduces
the yield. Circulating platelets possess histocompatibility antigens and
have a much less complex membrane than that found on the lymphocytes
which has been very well characterized. Studies by Phillips have demonstrated which components are on the surface of the plasma membrane.
The lack of DNA in the platelet would also result in a preparation that is
easier to work with than that obtained from lymphocytes or lymphoblasts.
These facts suggested to us that the use of platelets to solubilize histocompatibility antigens with 3M KCl may yield a soluble preparation that
is easier to work with and better characterized than that obtained from
lymphoblasts.

Platelets are generally isolated in the presence of a chelating
agent, such as GDTA, to prevent aggregation. It was of interest that if a
KCL extract was performed on platelets isolated in 10 mM GDTA, we could
not recover any soluble antigen activity in the KCl extract. In the rest of
these experiments three fractions were analyzed. The KCl extract, that
portion of the residue insoluble in 3 M KCl but soluble in isotonic sodium
chloride, and that which is insoluble in either solvent. The fact that, as
described above, a chelating agent could not be used, is of interest since
Oh, Pellegrino, and Reisfeld have also found that 3M KCl extraction of

lymphoblasts in the presence of EDTA causes an inhibition of the ability of 3 MKCl to extract antigen. The question then remains as to how to keep the platelets from aggregating. It turns out that in the platelets cyclic AMP levels are very important in preventing this aggregation. One way of doing this is to use phospho-diesterase inhibitors such as caffeine and theophylline. If caffeine is used, material that is soluble in KCl is obtained although it is not all that is present on the platelet. However, this compound does not always work in all individuals. I should stress that in all of these experiments, since we are looking at an allelomorphic system, it is necessary to use individual donors for each experiment. When activity was not found in the KCl extract, it was found in the sodium chloride extract as well as in the insoluble residue. We were plagued with this variability seen, using caffeing for some time, and then we tried theophylline and found that we recovered no activity in the sodium chloride extract but could consistently recover activity in the KCl extract, although not the total activity that was present on the platelet. This may reflect the fact that this compound by virtue of being more hydrophobic than caffeine, enters the platelet more easily. All of this suggested that cyclic AMP may be playing a role in the ability of KCl to solubilize antigen. To test this hypothesis, cyclic AMP levels were determined in platelets isolated under different conditions. Caffeine, although it did prevent aggregation, did not significantly raise the levels of cyclic AMP, whereas when platelets were prepared in theophylline, conditions under which we obtained antigen in the 3 M KCl extract, a significant elevation of cyclic AMP levels (37-67%) was observed. These results suggest that cyclic AMP levels are important in order for 3M KCl to solubilize antigens. One can speculate on how this works. However, it does appear that the antigen in the membrane may be a large aggregate, and in the presence of cyclic AMP and KCl dissociates to less aggregated forms. One of the earliest dis aggregation products would be that material which is soluble in isotonic sodium chloride but insoluble in potassium chloride. This then ultimately goes to the product which is soluble in 3 M KCl.

Which one of these dissociated states will be the proper state to observe in vivo biological activity such as transplantation associated phenomena remains to be seen. However, by testing fractions of different salt solubility we now have the means to determine how much these proteins can be dissociated before they lose this type of activity. At any rate, in this system it is not just an agent such as KCl which is causing solubilization, but other factors are involved.

Finally, I would like to talk about a third system of interest to immunologists that may not be amenable to solubilization. This is cell surface determinants that are associated with the mixed leukocyte reaction. To those who are unfamiliar with this, it is a phenomenon in which leukocytes from two unrelated individuals are taken and mixed together. They then stimulate each other to synthesize DNA and ultimately go on to division. As Dr. Bach has shown, a one-way reaction can be obtained by treating a stimulator cell with mitomycyn C so that it does not divide or synthesize DNA. In collaboration with Dr. Ranney, a very simple question was asked: Is this reaction only due to a recognition phenomenon, or is there more required of the stimulator cells? To answer this we decided to see what would happen if we treated the stimulated cell with a fixative. Fixation of the stimulator cells with either glutaraldehyde(2.5%) or lanthanum chloride (10^2 M) completely abolished their ability to stimulate. This suggested that in addition to recognition, an active process may be required of the stimulator cell in order for it to stimulate.

A number of individuals in cell biology have been discussing elements of the cytoskeleton and their ability to maintain membrane assymetry. Compounds such as colchicine and vinblastine or vincristine, which are thought to disrupt microtubules, have been demonstrated to alter the distribution of membrane components and affect processes such as agglutination of cells by Conconavalin A. We therefore investigated the effect of pretreating stimulator cells with colchicin and vinecristine on their ability to stimulate. These compounds resulted in a loss of stimulating activity. This suggests that a specific distribution of receptor components may be necessary for stimulation to occur. It thus appears that we have a system here where soluble products, when obtained, may not be sufficient enough to observe a biological activity. A certain configuration within the membrane as well as an active process may have to occur in order to be able to see an effect. What I have tried to do is describe the three systems, one of which is readily amenable to solubilization, one a little more difficult, and one in which other than getting the determinants into solution may not be solubilized and still show any biological activities. I think in looking at any of these antigens on the cell surface, whatever they be, one will probably have to think in terms of all of the above problems with respect to a particular system and not generalize in attempting to obtain soluble membrane components.

REISFELD: I am sure there must be some comments and questions.

UHR: A point of information. I am not clear why you stated that the antigens responsible for mixed leukocyte reactivity are not readily solubilized. Is it the inability in your experimental conditions to stimulate cells? Do you have an antibody to the antigens in question?

PINCUS: I am saying this: You probably can solubilize the determinants, but if one wants to take these materials and then test them in soluble form to see that they will mimic what a whole cell does--this may be a very different type of problem.

UHR: Of course, that is asking a lot more of the antigens.

PINCUS: Yes, but this may be necessary if one wants to look at a membrane-associated biological process isolated from the membrane.

McKHANN: How can you be sure that you don't alter the antigenicity of the sites by fixation?

PINCUS: We don't have an anti-serum that detects these particular components, but we can draw an analogy in Con-A agglutination of tumor cells where Sachs and others have shown that it is an active process and not just a recognition of lectin on the cell surface. There you can treat a cell with glutaraldehyde and abolish agglutinability, but the binding of Concanavalin A to the cell surface is not altered in a glutaraldehyde-treated cell. Since we do not have an anti-serum to the MLC determinants to perform a similar experiment, we did look at which was Con-A binding, and this was unaltered by treatment with either lanthanum chloride and glutaraldehyde, colchicine, or vincristine. Thus, these treatments may not alter antigenicity of the MLC determinants. However, it will ultimately take an antibody to these determinants to specifically answer your question.

BACH: I think that the evidence which you have given us does not really substantiate the conclusion that this might not be a simple recognition phenomenon. There is no question that the different antigens determined by the major histocompatibility system, referred to as H-2, behave differently. The antigens important in mixed culture are determined by the same region, and may very well have a different behavior on the cell surface in terms of turnover and other things, from the serologically defined antigens. In the sense they behave differently, not only with respect to the kind of fixation you have told us about, but to heat treatment, treatment with ultraviolet radiation, and various other things. I think we will get into this somewhat tomorrow, and I would not like to leave the session with the concept that this is more than a simple recognition system, necessarily.

PINCUS: Although we have no direct evidence, we can only draw by analogy to other systems where these reagents I have described have been used, and there, when compounds such as colchicine which increases membrane deformability and glutaraldehyde, which causes membranes to become more rigid and inhibits the active process of Con-A agglutination, I think one can come to a preliminary conclusion that there may be more than recognition for the MLC to occur. The exact answer will ultimately be determined when we can ascertain whether completely soluble products do the job.

PRICE: I wasn't clear. Did you try extraction of antigenic material from your membranes with three molar KCl?

PINCUS: Which system are you talking about?

PRICE: The histocompatibility antigens.

PINCUS: This is work that Ralph Reisfeld did and maybe he would like to comment.

REISFELD: Certainly it is difficult to isolate HL-A antigens from membranes by 3 MKCl extraction. As Jack Pincus mentioned, the yields are low. Most of the antigen can be found on these membranes in active form following extraction with hypertonic salt. We tried to explain this phenomenon by assuming that there were really profound structural differences on the membrane, whenever they are prepared by conventional methods. Actually, we have now considerable data to show that such changes occur and that they drastically affect our yields. I was somewhat amazed, and also pleased, about Jack Pincus' explanation of the mechanism of KCl extraction. I never dreamed it would be that elegant. Actually, we always considered KCl to act as a very mild chaotropic agent simply disordering water structure and thus overcoming the hydrophobic effects on the cell membrane.

MARCHESI: There are two interesting points about these areas you have covered. First of all, are you going to talk about this KCl business tomorrow as to what it might be doing?

REISFELD: Yes, I can talk about it any time.

MARCHESI: As I understand it now, Jack, if you have raised cyclic AMP levels, then you have the possibility of extracting material which is not extractable at lower levels. Is that the idea?

PINCUS: That is correct.

MARCHESI: What is the mechanism? What do you think it is due to?

PINCUS: One can only speculate. What is known in the platelet is that when cyclic AMP is increased, membrane bound protein kinase is activated and there are four specific membrane proteins, as shown by Boyse and coworkers, that are phosphoylated. Thus, one of the models that can be proposed is phosphorylation of cell membrane components which may cause some kind of dissociation by increasing the negative charge and thus some proteins are now extractable in hypertonic salt. Another possibility may be that cyclic AMP combines directly with a class of proteins on the membrane and by virtue of that causes disaggregation. There is also evidence for this type of mechanism occurring in the platelet.

MARCHESI: Do you feel, Ralph, that protease is not the mechanism?

REISFELD: Actually, more than feeling we have some data which was just recently published. I don't want to belabor the point, but it is pretty clear cut that protolytic activity <u>per se</u> is not the major driving force behind the solubilization that occurs with KCl.

MARCHESI: The other points I want to ask you about, it seems like a fascinating observation in your first slide that if you grow these lymphocytes at high density, they lose their glycoproteins. Is that the point you are making?

PINCUS: They were less agglutinable and the yields of glycoprotein we got were very poor. The one experiment we did not do in these studies was to measure lectin binding to see whether what you propose was the case, since we were only interested in finding a correlate between cell concentration and glycoprotein yield. However, what you propose is a possibility, since Nicholson has shown that the binding of ricin to cells is altered during the cell cycle. The only point I wanted to make is that in order to obtain reproduceable results, we had to be very careful of the density to which the cells were grown. Ralph has also shown this for extraction of HL-A antigen with 3M KCl which is effected by density. If one is using cultured cells, the density to which they are grown, is important in terms of the membrane components that you are going to get out.

LERNER: I think it is important to point out that when you grow lymphocytes to high density you synchronize them. The easiest way to synchronize lymphocytes is the G_1 phase of the cell cycle.

PINCUS: These were not synchronized in the sense that they were drawn from a mass population at low density.

LERNER: I think they were.

PINCUS: It may be. However, there is one thing I would like to ask you. When you synchronize, as you have just described, do you feel that the cells are in G_1 at that point?

LERNER: Between G_0 and G_1!

PINCUS: Differentiate.

LERNER: This is a semantic difference and for our purposes it is not important. But let me answer you operationally. If you would take L 1210 and grow them to 2.0×10^6 cells/ml and measure the rate of DNA synthesis, it would be about two percent the rate at mid-log.

BACH: By diluting them, do you actually see a synchronized population, an S phase and then down again?

LERNER: Absolutely, and we published this in the past. As I mentioned, the time it takes before the onset of DNA synthesis is the time of G_1. If at this point the density of cells is decreased by dilution into fresh medium, there is a lag period (G_1), followed by DNA synthesis (S). Thus, by definition the cells were arrested in the G_1 or G_0 phase of the cell cycle.

EDELMAN: Do you actually see cycles?

LERNER: It is not as nice as that. To my knowledge, no one has been able to get the multiple "step ladder" type cycle S that you can get with prokaryotic cells.

PINCUS: Dick, the only reason I asked this question is that cyclic AMP is higher during G_1, and I think in those conditions under which Reisfeld and Pellegrino have done the experiment they got the best antigen yield. This then would correlate with what I am saying is occurring in the platelet.

LERNER: I think there is no question about differences. I was just trying to answer Vince's question about the difference between cells at different densities. Have you quantitated the amount of cyclic AMP as a function of the phase of the cell cycle?

PINCUS: We never have done this.

LERNER: What about the L1210 cells?

PINCUS: No, we have not done that, but that is one of the things we should look into.

FERRONE: I would like to point out that the statement that the yield of soluble HL-A antigens from cultured human lymphoid cells depends on the phase of the growth cycle is not valid for all cell lines. For example, we obtained similar yields of HL-A 2 and 7 from the cell line DPMI 8866 throughout the growth cycle. The other two points I would like to make concern platelets as a source of soluble HL-A antigens. Michele Pellegrino in our laboratory could solubilize HL-A antigens from platelets, but there are two limitations: the first one is the relatively poor yield, while the second one is the lability of the antigenic activity: after one month of storage all the activity is gone.

PINCUS: In terms of your first point I think one would still have to perform a correlation between antigen yield and cyclic AMP levels to test this hypothesis. I should also have made the point that the conditions Pellegrino used to isolate platelets, namely 0.03 M adenosine, nearly doubles platelet cyclic AMP levels. I do agree with you that the amount of protein is very small; but on the other hand, a billion platelets are not a billion lymphocytes, and a billion platelets in total contain only on the average one milligram of protein. This may not be the way to go for large scale isolation. As far as we have found, after one month the activity also decreases and we are not sure what the answer is there either.

REISFELD: We should probably try to go on after this very interesting presentation and discussion and now hear about HL-A expression as related to β_2-microglobulin expression. There is presently considerable discussion as to whether there are molecules for an antigenically functioning unit on the cell surface and whether β_2-microglobulin is essential for or possibly modulates the antigen expression of HL-A determinants. In this regard Wernet will now present some of his studies.

WERNET:[*] Polypeptides have been found to be essential for HL-A antigenic activity (Reisfeld and Kahan, 1971). Since purification procedures most often involved enzymatic digestion, it was concluded that it could not be chemically proven that the first and second HL-A series represent separate molecular products. In recent studies several groups demonstrated that highly purified HL-A antigens consist of two polypeptide chains, and that the larger subunit carries the antigenic specificity and is non-covalently bound to a smaller subunit which was identified as β_2-microglobulin. Beside the LA and FOUR series antigens previous studies have suggested the existence of other serologically detectable HL-A series antigens (van Rood and Eermisse, 1968). Thorsby, et al., revealed in 1971 the AJ series which has been classified as third series antigen together with other five antigens known in this group so far.

To study these HL-A antigens the surface radioiodination of lymphocyte membranes as described by J. Uhr's laboratory has been

---

* This work was done in collaboration with C. Jersild, B. Dupont, J. Hansen, A. Svejgaard, L. Staub-Nielsen, R.A. Good, and H.G. Kunkel

employed on heterozygous and homozygous cells together with anti HL-A
sera directed against first, second, and third series antigens. Membranes were solubilized with 0.5% NP-40. On Table 2 the figures obtained
in the specific precipitations are presented. The homozygous membrane
(Ha) seems to contain about twice the amount of HL-A as the heterozygous
membranes. Although caution is indicated in quantitation of this method,
this proved to be a constant finding on other homozygous lymphocytes.

Figure 19 shows the iodination profiles on homozygous lymphocyte membranes which were obtained with antisera monospecific for the
first (A), the second (B), and the third series antigens (C).

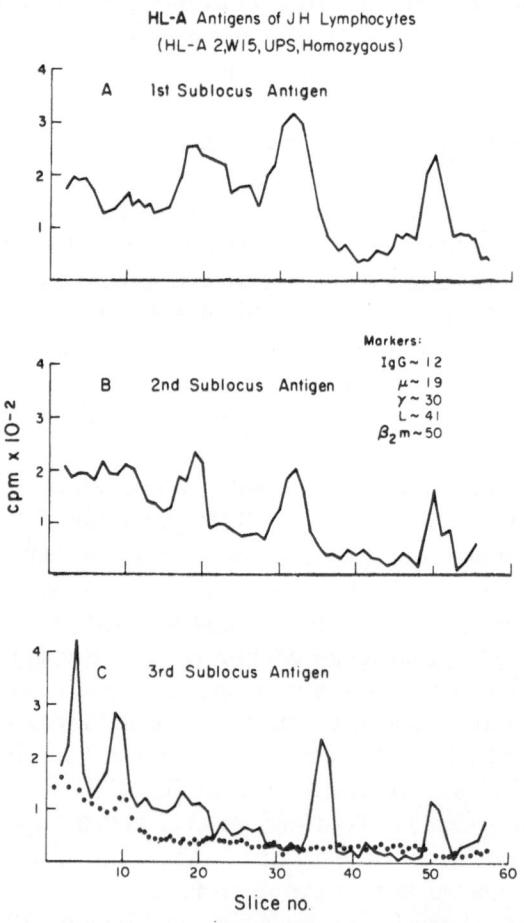

Figure 19: Surface radioiodination profiles of the HL-A antigens of the
first (A), second (B), and third (C) series. The third series major
polypeptide chain is clearly smaller. In all three gel patterns a peak of
HL-A dimers can be observed.

TABLE 2

Name	HL-A Type	PBL[1]	Total Acid Precipitable Radioactivity[2]	% by anti Ig	Percentage Radioactivity Immunoprecipitated by Anti HL-A Sera								By Anti-β_2Mic	By AB Serum
					HLA-2	W 15	27	W 22	315	170	AJ	UPS		
Hc	2, W15-UPS Homozygous!	5.6	2.6	5.50	2.47	2.93	-	-	0.52	-	-	2.48	-	0.15
Ro	2, W26, W10 UPS, W22-AJ	6	3.0	4.76	1.67	-	-	1.43	-	-	1.40	1.52	1.76	-
Se	W32, W26, MK W22-UPS	7.7	9.8	4.56	-	-	-	0.28	0.73	1.07	0.80	1.80	2.35	0.29
Cc	2, W19, 27-AJ W5-315	4.76	2.2	4.46	1.22	-	1.56	-	1.36	-	1.45	0.25	-	-
Be	3, 27-170, W15 UPS	9	6.0	4.95	0.43	0.89	0.88	-	-	2.29	0.77	1.48	1.39	-
Sti	2, 27, MK - - (neg. in AJ, 316 UPS, 170)	9.76	4.8	5.61	1.44	-	1.03	-	0.51	0.40	0.26	0.35	-	0.37

1 Number of peripheral blood lymphocytes ($\times 10^7$)

2 Counts per minute ($\times 10^6$)

          As already reported earlier (Wernet and Kunkel, 1973, Trans-
plant Proc. 5:1875), mainly three peaks are identified as the HL-A profile.
All three pictures show the presence of β_2-microglobulin at position No.50.
For the first and second series antigen the major polypeptide chains are at
position No. 33, which would mean as estimated molecular weight of 32,000
to 35,000 dalton. This peak for the third series antigens, however, is
clearly moving faster and it's molecular weight would be about 27,000
dalton only. The larger peaks at position 20 have approximate molecular
weight of 80,000 - 85,000 dalton and are candidates for HL-A dimers.
Indeed, there are two ways which indicate this strongly. Longer incubation
of isolated membranes in NP-40 solution decreases this peak considerably
and also the reduction of the specific HL-A-antiHL-A precipitors with 2-
mercaptoethanol results in disappearance of the high molecular weight
peak. This does not necessarily mean that an -S-S- bound association
exists. 2-mercaptoethanol can have different ways of acting on membrane
components.

          From these data, however, it would be possible to speculate
that HL-A on lymphocyte membranes can exist as dimers where two molecules of β_2-microglobulin would be the anchored mechanisms in the membrane and each one would carry one HL-A-specific polypeptide chain
which would be held together by an unknown binding mechanism. Perhaps
carbohydrate molecules could serve as mediators and stabilizers of the
constellation of SD and LD membrane determinants. The last figure
(Figure 20) demonstrates again the major polypeptide chain of the third
series antigen 315, which has a molecular weight of only 27,000 dalton
and does not cross-react with UPS, another third series antigen. To ensure the molecular size of the third series antigen, here a light chain
marker was used which in itself has a molecular weight of 27,000 dalton
(LT_r). This unusually large light chain was isolated from an IgG molecule which has activity against transferrin (Wernet, Kickhoefer,
Westerhausen, Clin. Res., 1972). Thus the present information indicates
(Thorsley, Transpl. Rev. 18, 1974) that the products of each HL-A allele
reside on different molecules with the major polypeptide chain of the first
and second series antigens at similar or identical molecular size of about
32,000 dalton in contrast to the major polypeptide chain of the third series
antigens which is about 27,000 daltons.

REISFELD: Thank you for this interesting presentation. I am literally
itching to get into a good discussion regarding the limitations of conclusions based on SDS-polyacrylamide gel patterns. I am sure there will be
an opportunity to do this later, but somehow I have to curb my enthusiasm
since we are running a little short of time. I would like very much to hear
more about the HL-A β_2-microglobulin story and I have thus asked Dave
Poulik to tell us more about it to get some kind of concept established about

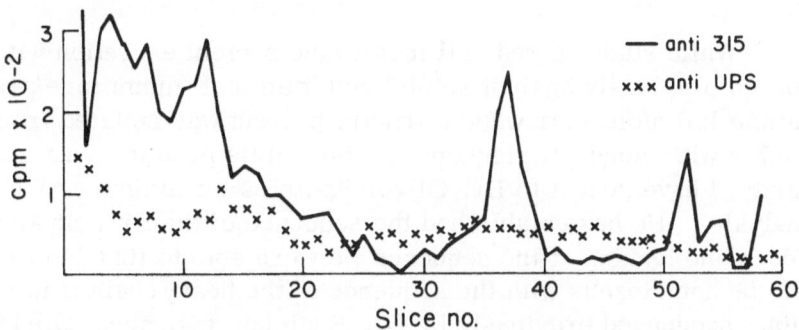

Figure 20. Surface iodination of the third series HL-A antigen 315. The major polypeptide chain is located in the position of marker L_{Tr} (molecular weight of 27,000 daltons).

these two cell surface markers.

POULIK: Listening to the morning group and to some of the members of this session who formulated the criteria which protein m a y qualify to be a membrane protein, β_2-microglobulin stands out as a pauper. It d o e s not contain carbohydrate, does not form micelles, its hydrophobic region is short, and the protein is probably too small to traverse the whole thickness of the membranes. In spite of all these shortcomings, it is probably the only membrane protein at the moment, which is characterized to the last detail from the physico-chemical and structural point of view.

β_2-microglobulin (β-2m) was isolated from urines of patients suffering with renal tubular deficiencies by Berggard and Bearn (J. B. C. 243:4095, 1968).   The protein was well characterized; i t s molecular weight was determined to be 11,600 daltons, its N -terminal amino acid to be isoleucine and it was shown to be devoid of carbohydrate and free S S - group.   The modes of this protein were determined in several biological guides.   Thus serum contains about 1. 6 - 2. 0 mg/liter; colostrum about twenty times this amount, and only trace amounts in normal urine.   Unfortunately, no specific biological function could be ascribed to this protein at that time.

While studying red cell membrane protein excretion with antisera prepared previously against solubilized r e d cell membrane protein components, a low molecular weight urinary protein was isolated from one of the patients with monocytic leukemia.   Since this protein w a s isolated in high purity, I have sent it to Dr. Oliver Smithies for amino a c i d s e - quence analysis.   He has established the sequence of the first 46 amino acid residues and looking at the sequence we were able to find 24 of these residues to be homologous with the sequence of the heavy chain of immuno - globulin En, sequenced previously by Dr. Edelman (Science, 175:18 7, 1972).   We have postulated a common evolutionary origin for this protein and immunoglobulins. Dr. Edelman will probably touch on this later. It has to be realized that in spite of this structural relatedness, no immuno - logical relation could be found.   Antisera prepared against this protein did not cross-react with any of the known immunoglobulin chains (gamma, alpha, mu, delta, epsilon, kappa, lambda, J-chain or secretory piece). Antisera against all these chains did not, likewise, cross-react with the β_2-microglobulin.   This intriguing protein was identified as β_2-m with an antiserum sent to me by Ingemar Berggard. Seven months after writing to Ingemar about these exciting results, Peterson, Bergaard, Cunningham, and Edelman reported the complete sequence of β_2-m (Proc. Nat. Acad. Sci. 69:1967, 1972).   They have c o r r o b o r a t e d our previous results as far as the degree of homology and sequence analysis have established, the h i g h e s t degree of h o m o l o g y exists with the C_H3 d o m a i n of the IgG heavy chain.   They proposed that the β_2-m is probably the f r e e

C_H3 domain which may contain all the effector functions of this part of the immunoglobulin molecule. Dr. Cunningham may add perhaps more to this part of the story.

These findings stimulated a great amount of activity to find those cells which manufacture β_2-m, using immunofluorescent technique. Mrs. Motwoni has shown that β_2-m is present on lymphocytes isolated from the blood of patients with chronic lymphocytic leukemia (presumably B-cells) and later on T-cells isolated from fetal and adult thymuses. Fanger and Bernia have shown that PHA stimulated lymphocytes secrete (shed) more β_2-m than unstimulated lymphocytes. In collaboration with Dr. Hutteroth, Litkin and Cleve we have studied the production of β_2-m and gamma globulins in long-term culture lymphocytes. The finding was that these cells produce in a unit time per 0.5×10^6 cells relatively equal amounts of β_2-m , whereas the same cells under identical experimental conditions produce widely different amounts of gamma globulin. Furthermore, we could show that the surface expression of these two different molecules was modulated independently of each other. Consequently, these two proteins are different gene controls. In collaboration with Dr. Arthur Bloom we have studied an additional 19 long-term lymphoblastoid cells and their supernates, isolated the shed β_2-m from the culture, and demonstrated by immunological and electrophoric methods that this β_2-m is the same as found in the sera and the urines of kidney transplanted patients.

Antisera against β_2-m were shown in my laboratory to be lymphocytotoxic even in high dilutions, and this property could be completely abolished by absorption with purified β_2-m . Immunofluorescent studies of tissue culture cells derived from a variety of organs, tumors, etc., showed β_2-m to be also present on the membrane of these cells. Consequently, it became obvious that probably all cells containing nuclei bear β_2-m on their surfaces. A detailed study of disrupted cell membranes of blood cells, e.g., red cells, monocytes, platelets, etc., by a Swedish group demonstrated that β_2-m could be found in/or on these membranes, except those of mature red cells. Our own studies on erythroblasts of patients in hemolytic crisis have demonstrated β_2-m on the membranes of these cells. Finally, Pontein and his collaborators demonstrated clearly that there are differences among cells with regard to the amount of β_2-m produced (6×10^5 cell/65 hrs). Thus lymphocytes isolated from Hodgkin's disease patients synthesized less β_2-m, and those isolated from various epithelial tumors three to four times more than normal lymphocytes.

The association of the presence of β_2-m and the presence of a nucleus in the same cell was explained by the discovery that the histocompatibility antigens and β_2-m form a complex molecule. The original discovery comes basically from two laboratories, that of Dr. Pressman in Buffalo, and Dr. Peterson in Uppsala. In Buffalo antisera were prepared in rabbits and primates which were able to detect a "common" antigenic

determinant on all papain-solubilized HL-A antigens derived from tissue
culture cells.   Dr. Strominger's group has found that the same antigens
could be separated into two fragments upon dissociation in SDS, urea, or
guanidine by gel filtration.   One of the fragments was of low molecular
weight (12,000 daltons), and one of higher molecular weight (about
33,000 daltons).   The low molecular weight fraction was shown to react
with the antiserum specific for the common antigenic determinant by
Pressman in subsequent work.   Furthermore, this group has isolated the
low molecular weight protein from the dissociated papain solubilized
HL-A antigens and determined the amino acid composition which was
similar to that of isolated β_2-m.   Partial amino acid sequence of the low
molecular weight protein isolated from lymphocyte tissue culture which
was exactly the same as that of β_2-m, as demonstrated by Nakamera,
Taginski, Appella, Poulik, and Pressman.   Furthermore, antisera prepared against the human β_2-m fully cross-reacted with the tissue culture
β_2-m and the β_2-m obtained by dissociation of the HLA-β_2-m complex.
It should be pointed out that higher molecular weight fraction (33,000
daltons) exhibited the appropriate HL-A specificity and did not react with
anti β_2-m antisera.

The Swedish group started with isolated spleen cells which
were digested with papain to solubilize the HL-A antigens. The HL-A antigens were separated on DEAE cellulose columns.   The peaks containing
HL-A activities were further fractionated on a gel filtration column. The
HL-A containing material eluted in the 40,000 - 50,000 dalton region. The
β_2-m was detected by radioimmunoassay in the region of 12,000 daltons.
These fractions were radiolabelled and used in studies with SDS-polyacrylamide gels and gel filtration on G-150 Sephadex. Immune complexes were
prepared between I^{125}-labeled HL-A antigens and specific HL-A antisera,
and the subjected to acid-urea polyacrylamide gel electrophoresis.   Two
radioactive peaks were obtained.   The fast migrating peak was β_2-m, and
one slower peak was isolated by the use of anti-HL-A antiserum.   Thus,
the highly purified antigens were shown to be composed of two types of
polypeptide chains.   The longer unit carried all the HL-A antigenic determinants, whereas the smaller one bore β_2-m. The two chains were shown
to be held together by hydrophobic bonds only.   Furthermore, isolated
HL-A carrying chains could be recombined with β_2-m isolated from urine.

In collaboration with Drs. Grey, Kubo, Colon, Strominger,
Creswell, Springer, and Turner (J. Exp. Med. 138:1608, 1973) we have
corroborated the results of these two groups using peroxidase-labeled
lymphocytes with specific anti HL-A and anti β_2-m antisera in order to
obtain immunoprecipates for SDS-acrylamide gel electrophoresis. These
studies demonstrated the presence of two different polypeptide chains in
the HL-A β_2-m complex.   It was necessary to show whether the same
complex is present on the intact, preferably living, cells. Two approaches

were chosen. In collaboration with Dr. Marilyn Bach, the recognition units for PH A stimulation, MLC reaction, and sheep red cell rosette formation were blocked with the specific anti β_2-m antisera. The second approach was to obtain evidence that the HL-A-β_2-m complex can be manipulated with specific antisera to form capping. In collaboration with Drs. Cepellini and Bernoco we demonstrated by the immunofluorescence technique that anti β_2-m antiserum induced capping and completely removed the HL-A antigenic structures, since they were not detectable when capped cells were examined with five specific anti HL-A antisera. It was also shown that IgG gamma globulin and the HL-A-β_2-m complex capped independently of each other (Poulik, et al., Science 182:1352, 1973). Our most recent collaboration with Dr. Painter and his group (Painter, Yasmeen, Assimeh, and Poulik, Immunol. Comm. 3:19, 1974), demonstrated that β_2-m can be detected by complement fixation assays and by the Cl inhibition assay. In the latter method, β_2-m was found to be as effective as human Fc fragments. β_2-m was also found to be cytophilic for guinea pig macrophages. Consequently, some of the "effector" functions postulated by Edelman were thus demonstrated.

The most recent experiments in collaboration with Reisfeld used β_2-m as a handle to isolate the HL-A complex, and ultimately HL-A antigens. In spite of the report by Peterson that HL-A β_2-m complex does not react with anti β_2-m antiserum, immunoabsorbents were prepared which show some promise for the purification of HL-A materials. Whether β_2-m is only fortuitously bound to HL-A polypeptide chains or has some definite function, for example, in the modulation of the receptor binding site, remains to be seen. One of the key areas for further work is the examination of genetic polymorphism of this protein. So far efforts in this laboratory in this regard have been fruitless. Search for a β_2-m in other species has included   the isolation of dog β_2-m, which was sequenced by Smithies and Poulik (Proc. Nat. Acad. Sci. 69:2914, 1972). Rabbit and bovine β_2-m are being prepared for similar analyses. The presence of high molecular weight complexes of HL-A-β_2-m were recently demonstrated in human whey and urine. These materials are potential sources for the isolation of HL-A antigens.

WERNET: Antisera with specificity for β_2-microglobulin have been shown to inhibit mixed leukocyte culture reactions between allogeneic cells (M. Bach, et al., Science, 1973). In this regard we were able to confirm this observation. In addition, the following data on the inhibition of MLC reactivity by anti-immunoglobulin sera are very much related to this topic. In a large number of different anti-Ig sera tested it could indeed be demonstrated that such an inhibitory effect of certain sera does exist. When these sera were tested on radiolabelled lymphocyte membranes, they exhibited also an activity against β_2-microglobulin in addition to their anti-Ig specificity. Thus, surface radioiodination provides a powerful tool to test immunological monospecificity of antigen directed against membrane components.   Figure 21 shows an example of an anti-Ig "M" antiserum which blocks most allogeneic MLC combinations. At position 48, the β_2-m peak is identified.   This serum also reacts in agar diffusion with β_2-m a s tested later.

　　　　　　Figure 22 demonstrates the same finding for an anti-kappa-anti-serum, which was made in a rabbit by injecting isolated Bence-Jones-Kappa chains. One out of four anti Bence-Jones antisera contained additional activity against β_2-m. These data could at least in part explain the MLC inhibition described in the literature (review Greaves, Transplant Rev.). Anti-immunoglobulin sera which gave just their expected mono-specific precipitation pattern with lymphocyte membranes on SDS-acryla-

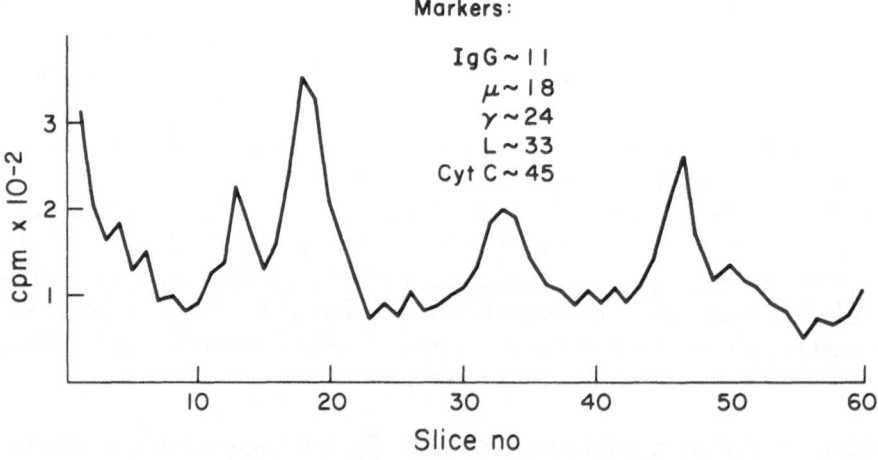

Figure 21: Surface radioiodination profile of peripheral blood lymphocyte membranes, solubilized with NP-40 and precipitated with an MLC-inhibitory anti-IgM antiserum. This antiserum is "contaminated" with antibody specificity against β_2 microglobulin; this can explain the MLC-inhibitory capacity of the antiserum. Blocking experiments with isolated Ig-M do not abolish MLC-inhibitory capacity.

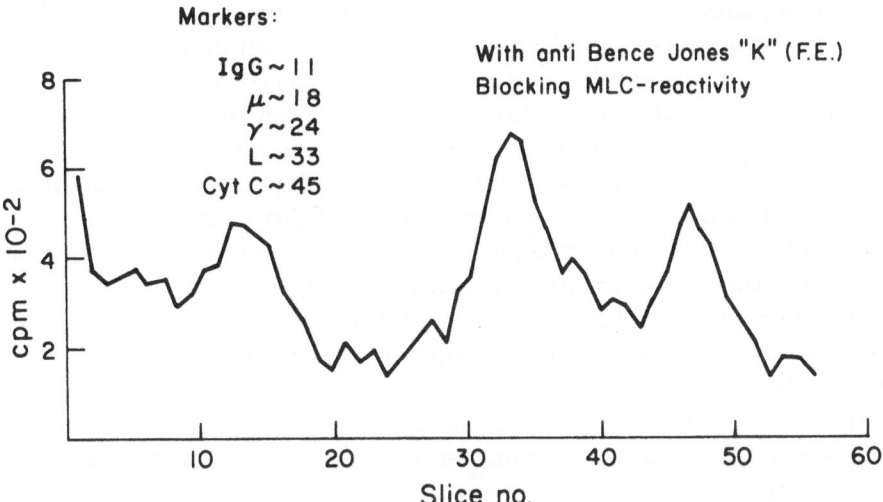

Figure 22: The same finding as in the previous figure, however, with an anti Bence-Jones-Kappa antiserum. MLC-inhibitory activity can be absorbed out with isolated β_2-microglobulin.

mide gels, did in no case inhibit any MLC-reactivity in our hands. Thus the receptor for MLC does not seem to be an immunoglobulin molecule. However, also β_2-m is unlikely to have specific receptor function, since so far no heterology has been observed. This would imply that β_2-m would be closely attached to specific receptor molecules and exist on the lymphocyte membrane not only in association with HL-A antigens.

REISFELD: Following this interesting presentation I am sorry to say that we have run out of time and I thus would like to briefly put into focus some of the problems involved in the solubilization of membrane components.

It is apparent from this far ranging and interesting discussion that some progress has been made in developing new approaches for the solubilization of some cell surface components and in gaining a better understanding of their biologic relevance and cell surface expression. It is clear, however, that presently available methods of solubilization are quite non-selective and thus leave much to be desired since they essentially create a "needle in the haystack" problem. In other words, although reasonable amounts of some cell surface components can be rendered soluble, there are always much larger amounts of contaminants more often than not comprising 99% or more of the solubilized material. This problem necessitates the application and often the development of highly efficient purification procedures. Although this has not been discussed in any detail during this afternoon's session, it can be stated that considerable progress has been made in this area largely due to the effective application of large-scale electrophoretic methods and the use of immuno- and lectin absorbents. For example, Dr. David in my laboratory has been able to utilize a number of specific lectin adsorbents in such a way as to effectively purify a rather complex tumor associated antigen, i.e., carcinoembryonic (CEA) antigen. The utilization of specific immunoadsorbents for the purification of HL-A antigens as discussed this afternoon provides another example of the successful application of this type of purification strategy.

Although it has been possible to isolate cell surface components from intact human red cell membranes in sufficient amounts to determine their chemical structure, this proved considerably more difficult in the case of lymphoid cells, most likely due to the difficulty in obtaining pure, intact membranes and also because of the vastly increased complexity of the surfaces of these cells. This particular problem led several investigators, including my colleagues and I, to search for soluble HL-A antigens in serum. Such antigens can indeed be detected, isolated in reasonable amounts, and purified extensively. Most likely these antigenic moieties represent shed cell surface products. These materials are not only highly useful for comparing their chemical composition and structure with that of identical components isolated, but also can be used to prepare sensitive immunological reagents suitable for mapping of the cell surface, to develop sensitive assays to detect soluble cell surface antigen in the serum and to assist in their purification. This approach thus offers constructive alternative to the isolation of soluble components directly from the surface of cells as it actually complements this effort.

SESSION III

MODULATION OF CELL SURFACE
STRUCTURE AND FUNCTION

Lipide artificial membrane models - Glycoprotein-lipid interactions - Phase modulation - Macromolecular modulation of lymphocyte surface receptors - Mitogens - Patching and capping - Microtubles and microfilaments - Gap junctions - β_2 microglobulin - Conconavalin A - Cytoplasmic cortisol receptors - Cell cycle membrane states - Cell-cell adhesion - Purification of tissue-specific cell-ligand activity.

SESSION III

MODULATION OF CELL SURFACE STRUCTURE AND FUNCTION

EDELMAN: Yesterday we heard a lot about structural features of the red blood cell, and we heard a lot about how difficult it was going to be to isolate clean components from all of these different cell surfaces. Today, I would like to turn our attention towards the function and dynamics of the cell surface, not that we want to ignore structure, but just because we don't really have a great deal of direct information on it. So I have asked the speakers each to spend about fifteen minutes or so on a variety of topics which I had hoped would represent some kind of ascending order, not of scale or intelligence, but of complexity. Looking at this list, you might say that Moscona is the most complex, and McConnell is the most profound, but that isn't really the way I wanted to arrange it.

In line with what was said yesterday, it seems to me that we have to start with some of the physical chemistry of membrane, and then work up the scale with a series of phenomena involving modulation of the receptor movement. Finally, we have to deal with specialized structures on the cell surface, and then with a series of biological approaches that I think are relevant to this meeting: Mitogenesis, the selection of mutants with interesting properties, the relation of genetics to cell surfaces, and finally that very fundamental subject, cell-cell recognition. So without any further comments, I will call on McConnell to start us off.

McCONNELL: I shall say a few words about lateral interaction in membranes and perpendicular transport through membranes. Consider two molecules, dielaidoyl phospatidylcholine (DEL) and dipalmityl-phosphotidyl choline (DPL). DEL melts at about 12 degrees. It undergoes a first-order endothermic phase transition in which the hydrocarbon chains melt and there is rapid lateral motion of the phospholipid molecules. DPL melts at about 42 degrees centigrade and undergoes the same type of phase transition. This phase transition in the case of DEL can be detected from the paramagnetic resonance of the label called tempo. Tempo dissolves in the fluid hydrophobic region of DEL and does not dissolve as well in the solid phase of this phospholipid. Thus, this phase transition can be detected by means of paramagnetic resonance.

If one takes a sample of DEL at a temperature below the transition and rapidly quenches it and carries out the freeze fracture process and takes a photograph of the replica, one sees an array of parallel bands which indicate a crystaline state of the phospholipid molecules. If one takes the same material above the melting temperature and rapidly quenches it, one obtains either jumbled bands or no bands at all, indicating that at the

ambient temperature from which the sample is quenched the lipid is in a
fluid state. The same thing is true with DPL. Therefore, in this freeze
fracture experiment one can distinguish whether or not the lipids are in a
fluid state or a solid state at the ambient temperature of quenching.

       A phase diagram of the type shown in Figure 23 was constructed
by Mr. Wu, using mixtures of these two lipids and the same spin label tempo that I mentioned before.

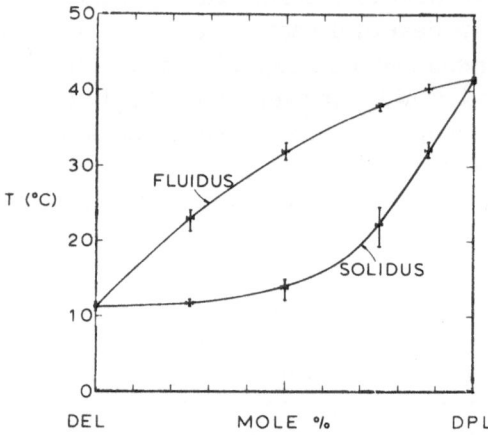

Fig. 23. Phase diagram describing lateral phase separations of binary
mixture of DEL and DPL.

       The idea of the experiment is that at one particular composition
of the two lipid molecules and reduces the temperature, one arrives at a
temperature t_n such that the solubility of the thin label in the lipid mixture
decreases at a different rate. The temperature t_n marks the onset of the
lateral phase separation. This rate of decrease changes again when all the
lipids are in a solid stage, at a temperature t_e. At intermediate temperatures, according to a phase diagram of this type there is a lateral phase
separation of the lipid molecules, one of which is rich in DPL melting lipid
and forms a solid domain; and another that is rich in DEL and forms the
fluid domain. Grant and Wu have carried out experiments in which this
phase diagram has been verified by rapid quenching and freeze fracturing.
For example, if one prepares a mixture of these two lipids in a fifty-fifty
proportion and selects a temperature which puts the sample right in the
middle of the phase diagram and then rapidly quenches the system, one obtains a freeze fracture pattern which is clearly a system of two domains.
roughly fifty percent solid, crystalline lipid, and the other the fluid domain.
The widths of the solid and fluid domans can be large, of the order of ten
or twenty thousand angstroms. The domain boundaries of this binary lipid

system are linear. This domain boundary shape is not characteristic for
all binary mixtures of lipids.

If one extracts glycophorin from the human erythrocyte, according to the method previously described by Marchesi and collaborators,
and reconstitutes glycophorin into this particular mixture of lipids, then
the phase diagram describing lateral phase separation of the lipids is not
appreciably perturbed at the concentration of glycophorin employed; this
concentration is of the same order of magnitude as it is in the erythrocyte.
As an example, if one takes this ternary mixture of these two lipids, plus
glycophorin, and selects a temperature and composition such that one is
right in the middle of the phase diagram and rapidly quenches the system,
one obtains an extraordinarily simple pattern. Fifty percent of the freeze
fracture surface consists of solid, crystalline lipids containing virtually no
glycophorin, and all the remaining glycophorin is present in the fluid phase.

There is apparently some tendency for the formation of dimers
and trimers of the glycophorin molecule. The glycophorin does not line up
along the domain boundaries between the fluid and solid phase. It is exclusively in the fluid phase. Unfortunately, I don't have these photographs
here, but they are in press in the Journal of Supra Molecular Structure
(W.K. Keeman, C.C. Grant, and H.M.McConnell). It is possible to grow
cells with relatively simple fatty acid compositions that approximate binary
mixtures of phospholipids. For example, Linden and Fox have obtained un-
saturated fatty acid auxotrophs of E. coli cells which are also deficient in
the β-oxidation of the fatty acids. In these E. coli cells the lipid phase behavior is essentially this type; isolated inner membranes can be characterized
with these spin labels as having a temperature t_n at which the lateral phase
separation of the phospholipids begins, and a temperature t_e at which the
lateral phase separation of the phospholipids ends.

Wolfgang Kleeman and I have carried out freeze fracture studies on these inner membranes, prepared from this double mutant of E.
coli and have seen, at least approximately, the same type of result I described for glycophorin, namely, that at high temperatures the membrane
particles in the system are in the fluid phase and are randomly distributed,
but at lower ambient temperatures the membrane particles are excluded
from the crystal domain that is present in equilibrium with the fluid domain.
Linden and Fox have made what I think is a most extraordinary observation
on the rate of β-galactoside and β-glucoside uptake into these coli cells, as
a function of temperature, in a situation where the membrane contains
essentially only two types of phospholipids. Specifically, in the case of
this E. coli auxotroph the membranes consist predominantly of lipids that
either contain two elaidic acid chains or contain one elaidic acid chain and
one saturated fatty acid chain. The result that they observe is that as the
temperature is lowered and the rate of active uptake of β-galactoside and
β-glucoside is assayed, then the rate of uptake decreases until the

temperature t_n is reached. When the temperature t_n is reached, there is a seemingly discontinuous <u>increase</u> in the rate of sugar uptake. This is a remarkable result because in terms of the phase diagram of this type when we reach this fluidous curve (temperature t_n), the amount of solid lipids in equilibrium with the fluid lipids is mathematically infinitesimal. It is only as you progress deeper in the phase diagram that one obtains more and more solid lipids in the system in the membrane. My thought as to the interpretation for this spectacular increase in transport at this tem-perature corresponding to the onset of the phase separation is that when two phases coexist having different areas per molecule, then the system has a high lateral compressibility. High lateral compressibility is asso-ciated with large density fluctuations. Both compressibility and density fluctuations could facilitate the insertion of the transport protein into or through a membrane and should also facilitate the contraction cycle of one or more components of the translation system.

There is only one difficulty with this argument, which is that the amount of solid lipid in equilibrium with the fluid lipids at the tempera-ture corresponding to the onset of phase separation is, as I said, extremely small. So how could a virtually infinitesimally small proportion of solid lipids have this spectacular effect on transport? The only conceivable way that this could happen, it seems to me, is that if the transport protein acted as a nucleation center for the crystallization of a "solid" lipid, so that as one approaches, enter, and crosses this part of the phase diagram, they surround these transport proteins and form a fluctuating environment of solid and fluid lipids. Thus, there is a local high lateral compressibility and density fluctuation. We have sought to demonstrate this effect by the following experiment. We have not carried out the experiment with the E. coli system because we don't have the isolated transport proteins, but we went back to the system of glycophorin. We attached a spin label co-valently to glycophorin and reincorporated this labeled glycophorin back into the binding mixture of the phospholipids. So now we have a system that I described previously, except that the glycophorin is spin-labeled and it is paramagnetic. The spin labels are attached relatively non-specifically and covalently. In addition, dipalmitotyl lecithin, DPL, was specifically labeled with C^{13} in the methyl region, choline head group region of DPL. The expectation was that if some very special phenomenon took place at the temperature corresponding to the onset of the lateral phase separation in that system, namely, the system of two phospholipids plus spin label glycophorin, something spectacular should happen at this temperature, corresponding to the onset of the phase separation, since we monitored the carbon nuclear 13 resonance.

The next slide shows the preliminary data obtained by Phillip Broulet:

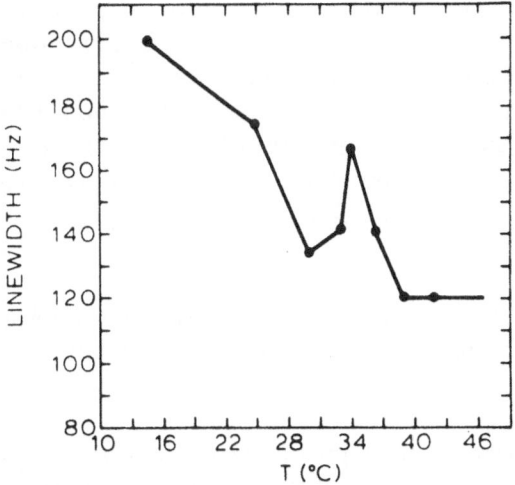

Fig. 24. Carbon-13 nuclear resonance line with N-CH₃ C^{13} labeled DPL in a binary mixture of DPL and DEL containing spin-labeled glycophorin.

This is a carbon 13 nuclear resonance liner width for the DPL molecule, labeled by carbon 13 in the polar head group in the presence of paramagnetic glycophorin. At the temperature according to the onset of the lateral phase separation there is a large increase in the nuclear resonance line width. What happens, I think, is clearly this. At high temperatures in the case of glycophorin inserted in this binary mixture of phospholipids where there is no solid lipid present, the molecules containing C^{13} undergo rapid lateral diffusion in the plane of the membrane and undergo short-lived collisions with the glycophorin molecule. At the characteristic temperature t_n at which the onset of the lateral phase separation begins the phospholipid molecules begin to stick to the glycophorin membranes, forming separate domains of phospholipids. This enhanced sticking increases the lifetime of the magnetic interaction between the C^{13} nuclei and the unpaired spin, or unpaired spins of glycophorin molecules. As the temperature goes somewhat below t_n there develops around the glycophorin molecule a kind of halo of phospholipids where the rate of exchange is less rapid. Therefore the distance of closest approach of the fluid lipids to the glycophorin is greater, and that accounts for the decrease in the carbon 13 nuclear resonance line-width at lower temperatures.

This halo of relatively solid lipids is to be distinguished from the situation which one would have if glycophorin were embedded in the total solid phase of the phospholipids because, as I mentioned before, we have direct freeze-temperature evidence that the solid lipids and fluid lipids coexist, and the glycophorin is exclusively in the fluid phase. So the conclusion is that when solid lipids and fluid lipids coexist, molecules such as glycophorin are "dressed up," they have a halo of lipids, which undoubtedly increases the dynamic, lateral compressability in the immediate vicinity of that membrane protein. I think this phenomenon accounts for the enhanced transport in the E. coli system that I mentioned previously. I think this experiment indicates a general class of magnetic resonance experiments that may facilitate the study of specific lateral interaction between proteins and lipids, and glycoproteins and glycolipids, and other components of cell membranes.

EDELMAN: Thank you very much. This is very intriguing. The subject is now open for discussion. Does anyone want to comment on angelic glycophorin?

REISFELD: I would like to know how critical the temperature changes are which affect your phase diagram. For instance, if you think of an in vivo situation where you have, for example, mammalian lymphocytes where obviously situations like this may occur, would you have this kind of lateral interaction and perpendicular transport when the cells are at 37°, or would there be a drastic change when the temperature would change to 35°C or less because at lower temperature there would be little correlation to a physiological system.

MC CONNELL: It seems to be certainly true in a number of bacterial cell system, especially mutants, that have simple fatty acid compositions, the onset of lateral phase separation, as I described, corresponds to the growth temperature. The work by Fred Fox and collaborators on animal cells using spin label biochemical assays and enriched fatty acid supplements seems to work on the onset of phase separations at or near the growth temperatures. There is nothing in what I said this morning that precludes adequate growth at lower temperatures. The growth may be slowed down for other reasons. It also is true that when all lipids are in the solid state then, of course, growth decreases in a spectacular way.

SONENBERG: I should like to speak to the issue of fluidity and its possible biological control, and for this I should like to show two slides. We have been studying a liver membrane system and the consequences of interaction with growth hormone. To get at the sampling of specific proteins in these biological membranes we have employed specific fluorescent ATP substrates. It could be a substrate as well for adenyl cyclase. Here is a polarization of fluorescence excitation spectrum. The lower curve is without growth hormones, i.e., just the liver membrane and the E-ATP, and the upper curve with growth hormone showing increased polarization.

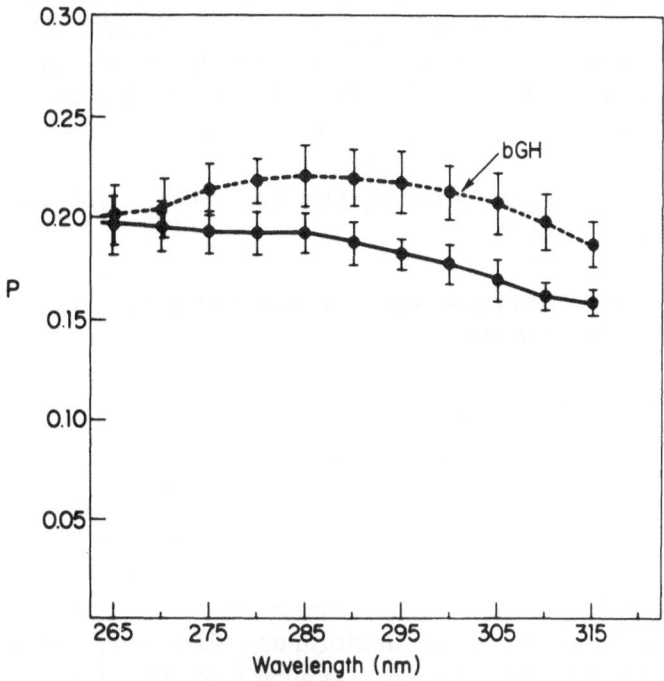

Fig. 25.   Fluorescence polarization of epsilon-ATP with l i v e r plasma
membranes in the presence and absence of 1. 2x10^{-12}MbGH.   The mixture
subjected to fluorescence polarization measurements contained l i v e r
p l a s m a membranes (87.1 ug protein/ml),   8. 2x10^{-6}M e p s i l o n - A T P,
3. 8x10^{-8}M GTP, 100 mM KCl,   5 mM MgCl$_2$ and 25 mM Tris-HCl, pH 7. 4
• - - • with bGH, • —— • without bGH.   Vertical b a r s represent s t a n d a rd
error.   Emission at 395 nm.   Emission bandwidth 9. 7 nm.   (Archives
of Biochemistry and Biophysics, 1974, in press.)

          Lifetime measurements have been performed with E - ATP.
There is a decrease in the lifetime of E-ATP in the presence of growth hormone. With the increase in polarization and decrease in lifetime, this would
be consistent with a decrease in the fluidity of the membrane.   However,
now having gotten these data, we are at a loss as to what this a l l means.
First of all, as you can see, the lifetime is not a single component on this
logarithmic scale. We think this represents in part multiple components
due to degradation of E-ATP,   since that is a satisfactory substrate f o r
ATP'ase.   In addition, we think we have more than one compartment even
though we are directing ourselves to two enzymes.   So in a way w e a r e

envious of your system where you can just label glycophorin and follow it. However, if you want to work with an intact membrane and sample specific proteins in situ, it becomes more difficult. We thought that the use of specific fluorescent substrates as probes might be an approach. We have a long way to go.

EDELMAN: I would just ask Marty a question. How do you separate out the scattering components?

SONENBERG: The scattering component contributes little in the range we are interested in. Its lifetime is very long.

EDELMAN: To the right?

SONENBERG: Yes.

EDELMAN: Okay.

MC CONNELL: I think I might make a comment on your remark concerning the various components of the membranes. The concentration of membrane particles for intrinsic protein in typical membranes is sufficiently high, I think, that all the lipids are somewhat sensitive to their presence. For that reason we can use quite non-specific lipid spin probes to detect the effects of hormones on cells with high sensitivity. For example, lipid fluidity in human erythrocytes undergoes a change at the level of one molecule prostaglandin per cell (P. Kury and H.M. McConnell, BBA). In fact, one can use changes in the lipid fluidity to detect new types of receptors in the cells (W. Huestis and H.M. McConnell, BBA).

EDELMAN: Could I ask a question concerning mammalian cells, for example? You are emphasizing the idea that there is phase separation and there certainly must be multiple examples in membranes as complex as those that we have been discussing. If you suspect that your phenomenon occurs in a cell, what would the halos look like? Could we have mixed halos? Is there any way of extending the idea from the simple model to cases related to physiological function?

MC CONNELL: I suspect that in many membranes these halos are chemically highly specific. In this particular situation that I have described, which is a reconstituted membrane containing glycophorin, the specificity of the halo is largely of a physical nature; but in intact animal cells it seems to me very likely that some halos will contain molecules such as specific glycolipids. In fact, Dr. Chris Grant and I have started preliminary experiments with Dr. Hakamori, using globiside and glycophorin, to use a similar effect to see whether or not there is a site around the glycophorin that is bonded to globiside.

BRANTON: I wonder if you could characterize a little bit more the nature of the halo at different points in the phase diagram mostly with respect to composition and mobility of lipids in this simple system. For example, at the onset of crystallization, do you perceive the halo as being primarily DEL or DPL? In the middle of the phase diagram, do you perceive the halo as being liquid DEL?

MC CONNELL: That is an excellent question. It is especially risky to answer it because one can do an experiment to test any conjecture that I might have. The halo that would be present at the temperature t_n would presumably be rich in the high melting lipid, that is to say, DPL. As one proceeds down through the phase diagram (lessening temperature), the halo, I suggest, will become richer in DEL. But I am not certain of this conclusion. I suspect that is true. However, the question can be answered experimentally by specific labeling of the different types of lipid molecules with C^{13}, and finding out which lipid C^{13} resonance is predominantly broadened. I suspect that in many cells the compositions of the halo do not change dramatically in modest temperature target. I do want to emphasize that the glycophorin molecules in the binary fracture are not in the bulk solid phase.

   If you look at the microphotograph you see the solid phase completely free of glycophorin; the glycophorin sitting there in the fluid phase surrounded by this halo that is not revealed by the electron microscope.

BRANTON: Isn't that an enigma? It acts as a nucleation site yet is not in the bulk solid region?

MC CONNELL: It is an apparent enigma, and that is why it is interesting.

CUATRECASAS: Is it still an enigma if you have a single or limited (two to three) layer of phospholipid which acts differently than the bulk at the point of the phase separation? Can you calculate the percentage of the phospholipid molecules that would be involved in direct contact with glycophorin as a single (or limited) layer, knowing the number or mass of the glycophorin, and making rough calculations of its lateral surface area, were it an ellipsoid? Is there enough phospholipid in the solid phase or in the altered phase to account for more than such a limited layer of phospholipid?

MC CONNELL: I think the idea of a <u>single</u> layer of phospholipids is quite naive, except perhaps when we are dealing with very specific interaction such as that between glycoproteins and glycolipids. I think you may be familiar with several problems in solid state physics where one domain had boundaries. Generally, in all these situations one had a gradation in properties between two limiting cases; so what I think actually happens is that right next to the intrinsic membrane protein, the lipids are quite rigid. Then at the next layer they are more fluid. At the next layer there is still more hydrocarbon chain fluidity and eventually this lateral fluidity matches exactly the fluidity of the "fluid" lipids. Thus, there is no single ring fluid lipid molecule, but there is a distribution of fluidity around the protein which has a certain gradient half-width; and I don't know what the size of that half-width is.

PINCUS: As you lower the temperature and reach the top of the curve that you have just described, do you begin to see a more ordered distribution of glycophorin in your binary system or does one have to go further down and get a greater of solid lipid before one sees a redistribution of protein.

MC CONNELL: The lateral distribution of the glycophorin molecule appears to be totally random in the fluid phase until the area exposed by the glyco-phorin is comparable to the area of the fluid lipid.

TOMKINS: Temperature is not a biological variable but protein conforma-tion is. Can conditions which alter the conformation of a membrane protein influence the extent of a surrounding lipid ?

MC CONNELL: I am not quite sure I really understand your question. I think that embodies really several questions. I believe that the receptor molecule of the membrane could very well modify these local lipids. I think that probably the strongest influence that I am sure exists on the lipids is one which is modulated, for example, in erythrocytes, by spectrin. I am sure that spectrin in erythrocytes has a profound effect on the fluidity of the lipids and probably their behavior around individual proteins. But that is all I am sure about.

TOMKINS: A conformational change in spectrin could alter the distribu-tion of lipids?

MC CONNELL: I am sure it will alter their state of fluidity because we have measured that. The state of fluidity in erythrocytes is affected by prostaglandins, can be affected by acetylcholine, and these effects can be blocked by a acetylation so that there is a strong coupling between fluidity, probably lipid phase equilibria, and the physiological state. For example, the level of cyclic AMP inside the cell indirectly affects the state of these lipids and might in turn directly affect the activity of proteins.

EDELMAN: Thank you. We started out yesterday with the idea of fluidity and the ideas on the distribution of receptors in the membrane. Then we turned to the discussion of the structural effects of interaction of protein under the membrane with proteins in the membrane. Now I am going to call upon two persons who have studied the dynamic aspects of a similar problem, a problem of modulation. We might use the term modulation to mean any influence on the movement or distribution of these proteins that alters their simple diffusion. We just spoke of one set of phenomena that might be called phase modulation, and now we are going to talk about something that might be called macromolecular modulation, particularly from submembranous structures.

It seems to me that one is not going to explain the phenomena of modulation we are going to hear about in any very simple way. What is going to be required, I think, is a whole assembly of structures, not even as simple as the one that Dan Branton mentioned yesterday, but rather a complex of proteins to modulate the behavior of particular receptors. I am going to call on Ichiro Yahara and Marty Raff to talk about aspects of this problem.

YAHARA: The subject I would like to talk about is the modulation of distribution and mobility of lymphocyte surface receptors by cellular structures other than receptors themselves. When one type of protein and one type of lipid-bilayer are given and the protein molecules are embedded in the lipid-bilayer, and in addition, if we know sufficiently the chemical and physical parameters of these molecules and interactions, we may be able to predict the distribution and mobility of protein in the lipid bilayer.

However, if the membrane is not isolated from other cellular structures and interacts not only with the aqueous medium but also with those cellular structures, those interactions must affect strongly the distribution and mobility of surface receptors. Of course, this idea requires a couple of critical assumptions, and the assumptions must be satisfied.

Assumption 1: Such surface proteins penetrate lipid layer or are attached to other components which extend to the opposite side of membrane.

Assumption 2: Some cellular structures exist just under the membrane so as to interact with membrane components. To prove these assumptions generally, experimental data are not sufficient at present. But I would like to discuss two key experiments from which we can derive some conclusions about essential features of the cell membrane and its dynamics.

In 1971, Taylor, Raff, and their associates found that binding of antibody against immunoglobulin (anti-Ig) to surface immunoglobulin (Ig) on B-lymphocytes induced redistribution of surface Ig. Their results suggest that the initial distribution of surface Ig is random. But after binding with divalent antibodies, the Ig-anti-Ig complexes form aggregates and subsequently result in global movement of the aggregates toward one pole of the cell. The first step of redistribution is called patch formation and the second is called cap formation.

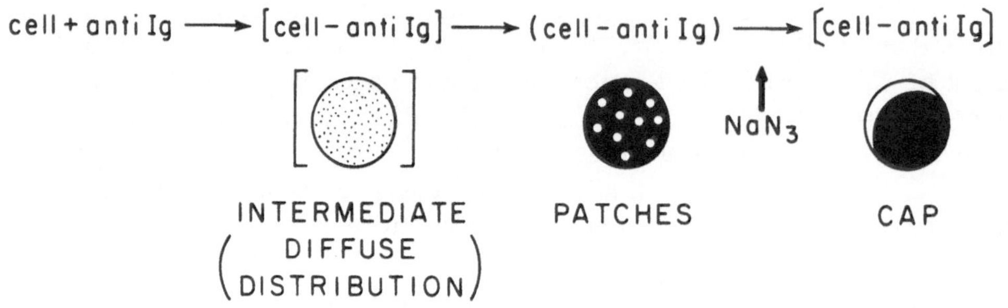

Figure 26:   A model for patch and cap formation.

Patch formation seems to be a rather simple phenomenon in terms of molecular interactions because this is very similar to the classical antigen-antibody reaction observed in the solution. Both require divalent antibody and multiple antigenic sites on the antigen. In certain cases of patch formation, both the low zone and the high zone effects of antibody concentrations have been observed, as is known in antigen-antibody reactions.

However, cap formation appears to be more complicated because this process is dependent upon cellular metabolism. Metabolic inhibitors such as NaN_3 inhibit cap formation. I think one of the most fundamental questions on cap formation is why or how aggregated receptor-ligand complexes move differently from free receptors. A cellular mechanism must exist which distinguishes aggregated receptors from free receptors.

It has been shown that cytochalasin B inhibited partially cap formation and it was suggested that microfilaments are involved in the mechanism of cap formation (Taylor, et al., 1971). However, other groups have reported that the drug did not prevent cap formation. Our recent experiments indicate that cap formation induced by anti-Ig is inhibited effectively (>80% inhibition) by 10μg/ml cytochalasin B in phosphate buffered saline. But at the same time, some samples of cytochalasin B inhibited cap formation only partially (10 to 40%) at the same concentrations. It seems important, therefore, to ascertain the purity of the cytochalasin B solutions used in these experiments.

Although the presence or absence of exogeneous glucose did not affect patch and cap formation and the effect of cytochalasin B, it cannot be excluded that cytochalasin B inhibits cap formation via binding to glucose transport sites or other membrane structures. However, the most likely mechanism is that microfilaments are involved in cap formation and cytochalasin B affects the function of microfilaments. If this is the case, the next question is how differently aggregated and free receptors are interacting with microfilaments. The question on the mechanism on cap formation remains to be solved. We have found that a plant lectin, Concanavalin A (Con A) inhibited both patch and cap formation induced by anti-Ig at 37°C. This Con A effect has been observed also on the redistribution of θ-antigens on mouse thymocytes of H-2 antigens on mouse lymphocytes and cultured mouse fibroblasts and of Con A receptors themselves. Therefore, binding of Con A did not result in patch and cap formation at 37°C. As shown in Fig. 27, the effect of Con A is dose-dependent and is effective as low as 5 μl/ml.

The inhibitory effect of Con A on redistribution of receptors is dependent also on the valence of Con A molecules. Native Con A is a tetramer of identical subunits at physiological pH's. But a chemically modified Con A, succinyl-Con A, is a dimer at the same pH and possesses the same affinity for the specific sugar, β-methyl-D-mannoside as that of Con A. However, succinyl Con A does not show any inhibition of patch and cap formation. In contrast to our results, it has been observed by Unanue, Karnovsky, and their associates that Con A did induce redistribution of its receptors. Their results and ours appeared to be contradictory. However, after testing the various conditions where incubations of cells with Con A were performed, we have found that the temperature at which Con A bound to cells affected the ability of cells to induce cap formation. Table 3 shows the influence of low temperatures on the inhibitory effect of Con A.

Table 3
Cap Formation Induced by fl-Con A (100μg/ml)

Experiment	Binding	Incubation	Cap forming Cells (%)
I	4°C	37°C	45
II	37°C	37°C	2

When lymphocytes were incubated with 100 μg/ml fluorescein-labeled Con A (fl-Con A) at 4°C, washed to remove unbound fl-Con A, and were incubated at 37°C, 45% of the cells showed caps (Experiment I). In contrast, when binding was performed at 37°C using the same concentration of fl-Con A, only 2% showed caps (Experiment II).

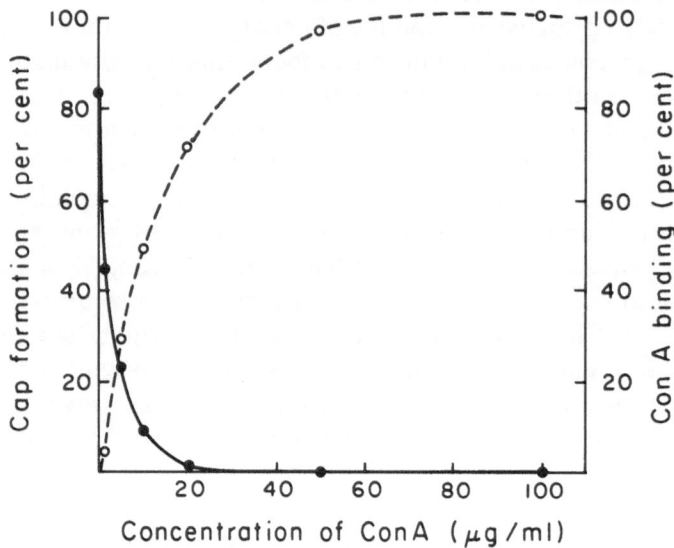

Figure 27:   Effect of Con A on cap formation (solid circles), and binding of Con A to cells (open circles).

Therefore, Con A exhibited two antagonistic functions with respect to its effects on redistribution of receptors depending on the incubation temperature. The question is: How does the low temperature affect the antagonistic activities of Con A? Three possibilities have been considered and examined.

(1) The molecular structure or subunit structure of Con A might be affected.

(2) The availability or arrangement of Con A receptor on cell surface may be changed.

(3) Cellular structures which are responsible for restriction of receptor mobility may be affected.

Possibilities (1) and (2) could not be neglected because it has been shown that low temperatures affect the equilibrium between tetramer and dimer of Con A subunits and dimeric Con A does not inhibit cap formation. In addition, it has been shown that the number of Con A molecules bound per cell was decreased by low temperatures. However, the results of various experiments indicate that possibilities (1) and (2) are not sufficient to explain the difference observed at the two different temperatures. Therefore, it seems that possibility (3), low temperature-sensitive structures might be involved. Microtubules are among the most likely candidates because of their temperature-sensitive structures. If so, microtubule-dissociating agents such as colchicine may affect surface phenomena induced by ConA similarly to the low temperature effect. We have found that this is the case.

Table 4 shows that microtubule-dissociating agents suppress the inhibitory effect of Con A on cap formation induced by fl-anti-Ig.

Table 4

Effect of Anti-mitotic Drugs on the Inhibition by ConA
of Patch and Cap Formation

Inhibitor	Drug	Caps (%) with fl-anti-Immunoglobulin
--	--	$80 + 11$
--	10^{-4}M Colchicine	86 ∓ 6
ConA	--	1 ± 1
ConA	10^{-4}M Colchicine	30 ± 8
ConA	10^{-4}M Colcemid	20
ConA	10^{-4}M Vinblastine	51 ± 6
ConA	10^{-4}M Vincristine	15
ConA	10^{-3}M Podophyllotoxin	10
ConA	5×10^{-4}M Lumicolchicine	1
ConA	10^{-4}M Griseofulvin	2

Preincubation of cells with colchicine reversed the ConA effect and allowed cells to form caps. Other microtubule-dissociating drugs, colcemid, vinblastine, vincristine, and podophylotoxin showed similar effects. An isomer of colchicine, lumicolchicine which has been known not to dissociate microtubules showed no effect. An anti-mitotic drug, but not a microtubule-dissociating agent, griseofulvin, did not suppress the ConA effect.
ConA itself can induce cap formation under conditions (Fig. 28) where the inhibitory ConA effect on redistribution of receptors is suppressed. In addition, our electron microscopic observations on ultrathin sections and ghost-membranes of cells incubated with ferritin-labeled ConA indicate that patch formation was induced on colchicine-treated cells but not on untreated cells. From the above observations, it was concluded that microtubules or related structures are involved in restriction of mobility

of receptors by ConA. Particularly, it must be noted that the regulation of receptor mobility by cooperative action of ConA and microtubules occurs at the individual receptor level. Now, it became clear to some extent that both microtubules and microfilaments are involved in modulation of distribution and mobility of surface receptors. The next question is whether these two cellular structures are acting cooperatively or not. The effects of colchicine and cytochalasin B on receptor mobility and the ConA effect on it are summarized in Table 5:

Table 5

Effect of Colchicine and Cytochalasin B on Receptor Mobility

	Patch formation	Cap formation
--	+	+
Colchicine	+	+
Cytochalasin B	+	-
Colchicine + Cytochalasin B	+	-
ConA	-	-
ConA + Colchicine	+	+
ConA + Cytochalasin B	-	-
ConA + Colchicine + Cytochalasin B	+	-

These results indicate that colchicine and cytochalasin B affect receptor mobility independently.

However, the following experiments suggest that microtubules may control or regulate the function of microfilaments with regard to their roles in receptor movement and cell movement. Using antigen-derivatized nylon fibers such as Dnp-BSA fibers, it has been shown that specific antigen binding lymphocytes were collected on the fibers. When these fiber-bound cells were kept in the medium at low temperature, the shapes of cells were mostly round. But when the incubation temperature was raised to 37°C, shape changes of fiber-bound cells took place within 30 minutes. The morphology of cells was different from cell to cell. Microcinematographic analyses indicated that the morphological changes are associated with cell

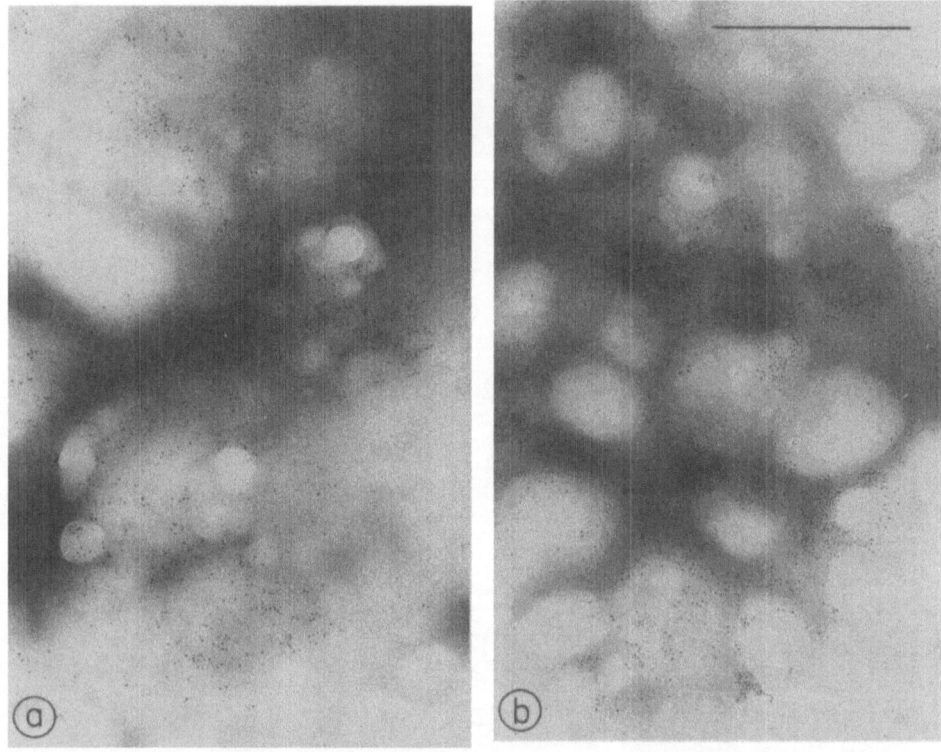

Fig. 28. Ultrathin section incubated with Ferritin-labelled Con A. Patch formation was induced on colchicine treated cells (b) but not untreated cells (a).

movement although these cells could not translocate along the fibers. It is interesting to note that only B-lymphocytes showed morphological changes. The shapes of T-lymphocytes were observed to be round during the entire incubation. Surface Ig molecules exist diffusely on these moving B-lymphocytes. When anti-Ig was added to these moving B-cells, cap formation was induced, the direction being a l w a y s opposite to the point o f attachment to the fibers (Fig. 29 a-c). However, when fiber-bound cells were preincubated with colchicine or vinblastine, morphological changes of the cells were inhibited. On these round cells, cap formation was still induced by anti-Ig, but the direction of capping was random (Fig. 29 d-f).

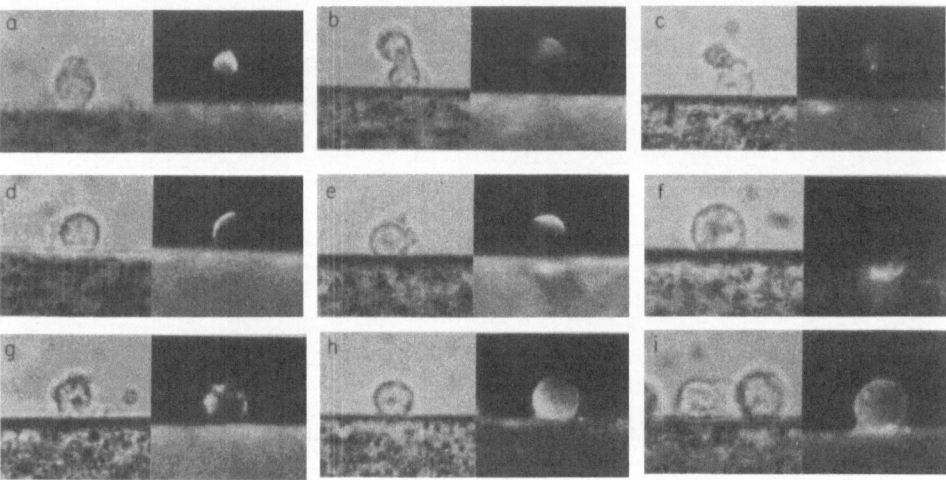

Fig. 29. Various types of fiber-bound cells labeled with fl-anti-Ig.
DNP-BSA fibers were used for (a) to (h), and ConA fibers were used for
(i).

Therefore, it is very likely that treatment by colchicine modulated the action of contractile proteins, probably microfilaments. Similarly, it has been shown by other groups using fibroblasts that colchicine-treatment abolished the directional movement of cells and replaced it with random movement. When we used Con A-derivatized fibers instead of antigen-fibers, morphological changes or fiber-bound cells and induction of patch and cap formation by anti-Ig were both inhibited. This Con A-fiber effect was partially reversed by α-methyl-D-mannoside without detaching the cells from the fibers. Moreover, cap formation can be induced by anti-Ig on colchicine-treated cells bound on Con A fibers. The directions of caps were again random with respect to the point of attachment to the fiber. The results indicate that local binding of Con A molecules to cells is sufficient to inhibit cell movement and patch and cap formation when microtubules are left intact. The involvement of microtubules and microfilaments in control of receptor mobility and cell movement is summarized in Table 6.

Table 6
INVOLVEMENT OF MICROTUBULES AND MICROFILAMENTS
IN CONTROL OF RECEPTOR MOBILITY AND CELL MOVEMENT

	Micro-tubules	Micro-filaments
Patch formation	-	-
Cap formation	-	+
Restriction of receptor mobility by Con A	+	-
Morphological change	+	+

Patch formation is independent of both structures. Cap formation depends upon function of microfilaments but not on that of microtubules. Restriction of receptor mobility by Con A depends upon microtubules. Morphological changes and cell movement as well as the direction of cap formation are dependent upon both structures.

Finally, I would like to summarize conclusions derived from our observations.

1) The cooperative actions of cross-linking of Con A receptors by Con A and microtubules cause restriction of receptor mobility.

2) Local binding of Con A to the cell surface is sufficient to inhibit patch and cap formation. Some structure that will transmit local structural changes of receptors or related molecules to the rest of the cell surface is required. We think this structure is the microtubule.

EDELMAN: Because we are running a little late, we are going right onto Marty Raff.

RAFF: I will deal briefly with two problems: (1) the interaction between membrane macromolecules and cytoplasmic components; and (2) the relationship of surface antigens and receptors to intramembraneous particles visualized by freeze-fracture. In both cases, I will be talking about mouse lymphocytes.

I think from the beginning, it was clear that capping probably involved contractile cytoplasmic components interacting with membrane proteins. Studies with cytochalasin B have shown partial, but reproduceable and reversible, inhibition of Ig capping on B lymphocytes. Since these studies were done in simple buffered saline without serum, glucose, or amino acids, transport effects seemed an unlikely explanation, and thus suggested that actin-like microfilaments might be involved in capping. On the other hand, microtubule-dissociating (M-D) drugs do not inhibit capping, making it clear that microtubules do not play an essential role. However, recently Stefanello dePetris has shown that when cytochalasin B and M-D drugs are used together, they completely inhibit Ig capping, and when vinblastine-treated cells are capped with anti-Ig, cytochalasin B reversed the caps in many of the cells. Thus it appears that cytochalasin B-sensitive structures, probably microfilaments, and vinblastine-sensitive structures, probably microtubules, somehow can synergize in the capping process. In addition, microfilaments appear to play a role in holding the cap together once it is formed. Since dePetris has demonstrated heavy theramyosin-binding microfilaments directly under the lymphocyte plasma membrane, it seems a reasonable working hypothesis that they can interact with membrane components and possibly also with microtubules. The latter may serve as a framework anchoring the filaments in some way.

Our results with ConA are quite different from those of Yahara and Edelman. In the first place, we have found that ConA caps its own receptors on 30-60% of spleen cells, even when the cells have been preworked and the reaction carried out at 37°C from the start. The majority of cells that cap are B lymphocytes while very few thymocytes or spleen T cells cap with ConA. Although M-D drugs enhance ConA capping to some extent, the effect is usually quite small. ConA capping is different from Ig capping in several respects: (1) there is very little patching with ConA and smooth rings convert to smooth caps; (2) there is relatively little pinocytosis induced by ConA; and (3) de Petris has shown that ConA capping is completely inhibited and reversed by cytochalasin B alone. In addition, dePetris has recently demonstrated that ConA caps Ig on the B cell surface. He has shown this using fluorescent-labeled ConA and rhodamine-labeled anti-Ig and also by immuno-ferritin electron-microscopy. Interestingly, Ig capping induced by ConA has the characteristic

of ConA capping in general, that is, there is very little patching or pinocytosis and it is completely inhibited and reversed by cytochalasin B.
Thus, our interpretation of why ConA inhibits the redistribution of Ig induced by anti-Ig is that the ConA cross-links Ig to a variety of other ConA
bonding glycoproteins on the membrane, thus preventing their clustering.
For some reason, ConA is a less effective inducer of capping and pinocytosis than is anti-Ig, perhaps because of the heterogeneity of the receptors
to which it binds.

At the moment, the reasons for the differences between these
results and those of Yahara and Edelman are unknown. It is unlikely to be
due to differences in the ConA used as in preliminary experiments. ConA
obtained from Edelman has behaved very similarly to the other ConA preparations we have used. It seems probable that subtle differences in the
way the cells are handled will turn out to be the explanation.

Finally, I want to discuss briefly the relationship between surface receptors and antigens and intramembranous particles (IMP) in the
lymphocyte membrane. Unlike studies on the erythrocyte where there is
increasing (although still indirect) evidence that PHA, ConA, and viral
binding sites and the ABO antigens are associated with IMP, all the studies
to date on lymphocytes have failed to demonstrate such a relationship.
Thus, Karnovsky and his colleagues have found that capping of a variety
of lectin receptors and antigens failed to induce any discernable change in
the distribution of the IMPs and such studies have now been confirmed in
a number of different laboratories. Leon Wofsy and his colleagues have
done the most convincing experiment of this kind by putting a hapten covalently onto available amino groups on the surface of lymphocytes using a
diazonium reaction and capping the hapten determinants with anti-hapten
antibody. Despite the fact that they could show that they had capped all of
the ConA receptors and all surface Ig (and thus probably the great majority
of exposed proteins and glycoproteins) there was very little change in the
IMPs. Recently Dunward Lawson and I, in collaboration with Bernie Gilula
have approached this question in a different way. Taking advantage of the
recent observations of McIntyre, Gilula and Karnovsky, that high concentrations of glycerol will induce clustering of IMP in lymphocytes, we have
treated lymphocytes and fibroblasts with 30-50% glycerol and then fixed
them in gluteraldehyde and looked at the dissolution of ConA receptors on
the surface using ConA- ferritin and thin-section electron microscopy.
Despite the fact that there was striking clustering of the IMP in many of
the cells, no clustering of ConA-binding sites could be demonstrated.

Thus it seems that one should be cautious in extrapolating from
findings on erythrocyte membranes to membranes of nucleated cells. It
may be, for example, that in the latter case the majority of IMP does not
extend to the external surface of the membrane. I do not think that one can
conclude that the surface receptors and antigens that have been studied on

these cells do not extend through the membrane, just because they may not
be associated with IMP. Kai Simon and his colleagues has shown that the
viral envelope glycoproteins of Semliki Forest virus are not seen as particles when the envelope is fractured, even though they have some evidence
that they penetrate through the envelope.

EDELMAN: Thank you, Marty.

KARNOVSKY: I do not want to belabor details of disagreement between
our observations and those of Yahara and those of Raff: this would prove
confusing. In general, there seems to be a reasonable degree of consensus.
        I want to talk about another cell type, namely the human poly-
morphonuclear leucocyte (PMN) which shows some differences in behavior
from the lymphocyte, but which illustrates some principles which we think
apply to the lymphocyte as well. This work was done in collaboration with
Dr. Graeme Ryan. When PMN are labeled with ConA in the cold and are
then warmed up, the cells crawl around and caps form at the trailing edge
of the cell, as can be clearly demonstrated in cells crawling under a chemo-
tactic stimulus. In the presence of colchicine, movement and tail caps are
observed. If movement is inhibited with cytochalasin B, or by placing the
cells in low serum, capping then occurs in the form of a central cap over
the nucleus area. If colchicine is combined with cytochalasin B, or if cytochalasin B is combined with prior cooling, no movement and no capping
ensues.
        We conclude that movement is not required for capping, but de-
termines the final location of the cap, i.e., at the tail. Capping in the absence of all movement requires an intact colchicine-sensitive system.
However, in the presence of colchicine, capping can occur, but only in the
presence of cell movement. If the PMN is labeled with ConA at 37°C,
movement is inhibited and capping occurs slowly over the central nuclear
area. Treatment with colchicine leads to active movement and rapid capping at the tail. Prior cooling of the cell and labeling in the warm has the
same effect.
        These results with ConA treatment in the warm are similar to
those of Yahara and Edelman in the lymphocyte, as are the effects of colchicine in reversing the inhibition by ConA on capping. However, we
think the colchicine treatment allows for cell motility which then leads to
the capping. If motility is inhibited, then capping--ConA in the warm,
plus colchicine--will not occur. Almost identical results have been obtained recently by Dr. Unanue and myself on the mouse lymphocyte. As
Yahara and Edelman have reported, ConA in the warm inhibits the capping
of anti-Ig-Ig complexes. Colchicine overcomes this inhibition of capping.
However, we find that a high percentage of these capping cells are motile,
and indeed, if motility is inhibited (e.g., by cytochalasin), the colchicine

reversal of ConA inhibition of Ig capping is not obtained.

We conclude (1) that ConA inhibits capping and cell movement; (2) the colchicine reversal of this effect is related to movement and will not occur in the absence of such movement. We do not think that the colchicine reverses the inhibition of <u>patching</u> of Ig produced by ConA. The behavior of patching in these systems may prove more interesting than the effects on capping. The main point I want to make is that the possible influences of cell movement on these systems should not be neglected.

EDELMAN: Thank you.

Perhaps you will clear up one misunderstanding which is purely semantic. We feel the colchicine effect is on patching and also that the ConA effect is on patching and therefore on capping. Everything else you said stands.

Now, I really think that this does provide a basis for us to relate to yesterday's talks, particularly the kind of thing that Dan Branton mentioned, although it is an extrapolation from one cell type to another. In Dan's case, we had structure without function, and here we have function without structure, although we do have a set of inferred structures which seem to be reasonable.

Bearing in mind the differences so clearly exposed by Marty Raff and Morris Karnovsky, I think we can open up the discussion in terms of macromolecular modulations and ask the question how it relates to the kind of phase modulations that Hardin talked about. I will now open the floor for discussion and comment.

MARCHESI: All these experiments are negative, and this is disturbing. Nobody has found a correlation between proteins in the lymphocyte membrane and the intra membranous particles, but my question is who has actually looked with markers that are known to bind exclusively to integral proteins in lymphocytes?

EDELMAN: To what?

MARCHESI: Integral, that is proteins supposed to be inserted in the membrane.

EDELMAN: Marty, the question has been directed to you.

RAFF: I think it is probable that all of the receptors and antigens that have been studied to date that are proteins or glycoproteins are "integral". I don't know of an example of one that is known to be "peripheral". The question of whether they penetrate through the membrane or not is another question. I do not think one should use the terms "integral" and "peripheral" other than as operationally defined by Singer and Nicolson. The characteristics of lymphocyte surface antigens and receptors clearly have the characteristic of integral proteins.

MARCHESI: Yes, but I guess the real question is what did ConA bind it to? Do you know what is it binding to in these cells?

RAFF: No, but I think it is likely that it is binding to the majority of the surface exposed glycoproteins and possibly some glycolipids.

MARCHESI: How do you know that?

RAFF: The data that make me think that are mainly those of Allen and Crumpton which indicated that the majority of detergent extracted glyco-proteins from pig lymphocyte membrane binds to ConA affinity columns. Of course, that may not accurately reflect the situation in the intact mem-brane. I think the most compelling experiments are the haptenization ex-periments that I mentioned earlier--where even when all the haptenated macromolecules (which should represent the great majority of surface exposed proteins and was shown to include all the ConA binding molecules and all of the Ig), were capped, the particles did not move.

EDELMAN: Perhaps you can sharpen your point. Is it that perhaps the experiment was not really critical?

MARCHESI: I think the point of the experiment described of having the particles clump artificially but not bind ConA ferritin is really a rather inconclusive experiment, unless you show that the ConA is binding to gly-coprotein. It could just as easily be binding to glycolipids.

EDELMAN: I think that has been established. It may also bind to glyco-lipids, but I think you can also isolate glycoprotein and glycopeptides that have specificity for ConA.

MARCHESI: Marty said that is true, but it does not mean ConA adsorbed on the cells is on the glycoprotein.

EDELMAN: What is your second point?

MARCHESI: The second point has to do with what the drugs are acting on. Is it possible that colchicine, cytochalasin B, or the other agents are actually modulating interactions with the membrane and do not act on cyto-plasmic structures; are there any experiments to rule that out is what I am asking?

EDELMAN: Yes, there is one. It is known that both lumicolchicine and colchicine are to some extent incorporated into membranes. Yet lumicolchicine has no effect whatever on the modulation effect. Cytochalasin, as Marty pointed out, is much more difficult to draw conclusions about. Pharmacological experiments are always open to criticism in the absence of complete structural evidence. Nonetheless, I think that the specificity of colchicine and its pharmacological analogues is extremely compelling. I would like to make one comment about your point related to intramembranous particles. Marty Raff only mentioned one kind of experiment that he and Bernie Gilula did. Yahara has looked at intramembranous particles under all the conditions he described, and the interesting thing is that in no case has he ever observed any redistribution of those particles.

MARCHESI: One other point. Dan knows as well as I do that it is sometimes very difficult to fracture where there is a large clump of particles, so it is possible that there is a uniform error. Finally, there is a possibility that there is only a small population of particles, perhaps that on the outer surface, which have glycoproteins.

EDELMAN: Now, I have a cluster of hands.

GILULA: We had several reasons for doing these experiments. One, work done on red cell ghosts has correlated movement of intramembranous particles concomitant with redistribution of receptor. We obviously wanted to look at the lymphocytes with every possible means of redistribution, both on the surface and within the plane of the membrane. Also, the experiments that were done by Unanue and Karnovsky indicated that there was absolutely no redistribution within the plane of the membrane concomitant with receptor redistribution. So we used the glycerol techniques as a way of trying to get redistribution within the plane, and then asked the other side of the coin, do we get redistribution of the surface?
        I think the important reason for doing the experiments were to find out whether or not in fact particles within intact lymphocytes have the kind of mobility that you see in red cells. Clearly, we have not yet found them.

MARCHESI: I was talking about the experiment.

GILULA: I hope nobody could infer you could not rearrange particles within the plane of the membrane concomitant with surface redistribution. We are saying that to date we have not found it.

KARNOVSKY: I think Dr. Marchesi's point is well taken. Dr. Unanue and I have not published our data in detail as yet because of negative results in regard to patching of particles, and it should be treated with caution, especially in the face of possible sampling error which is inherent in these sorts of experiment. All I can say is that we have searched for patching of particles in situations where we have extensive patching of surface entities: Ig, H2, ALG, theta, and ConA and PHA receptors, even patching several of these on the same cell. As yet no change in distribution of intramembranous particles has been found. We were particularly interested in patching and capping of PHA on the surface, as Dr. Marchesi has reported, that in the human red cell ghost PHA surface binding can be associated with the particles; and as Low has reported, capping of surface bound PHA yielding capping of intramembranous particles. We find no change in particle distribution with PHA.

Maybe we havn't looked with the requisite ligands, maybe there are technical problems such as sampling error, as Dr. Marchesi suggests, or maybe Dr. Raff is correct--the particles are not reflected at the surface.

MC CONNELL: I have a question, not a contribution. This morning the red cells have been contrasted with lymphocytes with respect to particle distribution, and in all the experiments that I am familiar with in the red cell in which particle redistributions have been observed, these cells have been subjected to really quite brutal treatment. We have looked very hard for a redistribution of the particles of the red cell that might be associated with function. We have never observed any such redistribution in particles. Joe Murdoch has shown the treatment of red blood cells with cytochalasin B enhances their susceptibility to complement mediated immune lysis suggesting a redistribution of antigens on the surface. But there again, we can detect no change whatsoever in the particle distribution. (Grant, Murdack, McConnell, unpublished.)

So it is not clear to me that there is such a big contrast between the red blood cells problems and the problems that are being discussed here with respect to the lymphocytes.

EDELMAN: I would just like to say I don't think we should get into a semantic controversy. I think Vince's point is well-taken in the sense that all these models, which I believe are tremendously unifying, and very useful for this meeting, remain to be proved at the level of structure in every case.

On the other hand, the semantic issue is what you call a particle, how you did the experiments, and all of that. I don't think how we could possibly solve it here. Dr. Hakamori had a point, and then Marty, and then we have to go on.

HAKAMORI: Just one comment. I have not really studied the morphological

relation to the receptor site interaction with macromolecular ligands, but
Dr. Galneberg and I have studied the interaction of glycoprotein and glycolipids with three reagents. One galacto-oxidase, two, recinus communis
lectin, and antibodies against galactoprotein(s). For example, galacto-
oxidase can pick up one galactoprotein, a high molecular protein, and the
major interacting point with galactose oxidase. But the reaction with
galactose oxidase is not easily blocked by ricinus communis lectin depending on the quantity of the lectin applied. Also, some of the glycolipids with
a galactose terminal are not labeled by galactose oxidase but can be inter-
acted with ricinus lectin. So these three different reagents, which all
should be directed to galactose residues, do not show any paralelisms in
their reactivities.

EDELMAN: Thank you. There is a question as to how any study of a
mechanism can be devised that would tell the difference between aggregated receptors and individual receptors.

RAFF: There is no evidence to suggest that it ever happens in the membrane, so we are talking purely theoretically; but there is certainly a
precedent for a protein having to aggregate before it can carry out a physiological function. For example, Ig has to aggregate before it can bind
the first component of complement.

EDELMAN: The point I was making is that despite the fact that we do not
have the solid structural evidence we would like among these structures,
for the first time we do have evidence of an assembly of proteins that
might have some specificity related to cell function, to cell movement,
and possibly to signals from the outside to the inside of the cell. Before
we get to functions of that type, I thought it would be a good idea to have
Bernie Gilula talking about an intriguing system. It is obviously of great
significance in the relationship of one cell to another, and it certainly
will have pertinence in making more real our models of how the cell surface
works. Bernie, I turn it over to you, and after your talk, we will have
discussion and a break.

GILULA: At this time I would like to talk about a highly specialized membrane. We have been studying a specialized region of membrane which has
a high degree of structural order, and the structural elements of this membrane can be observed with several different techniques.
    The structural elements of this membrane can be observed
with conventional thin sections, negative staining, and with freeze-fracturing. The elements of this membrane, which contain freeze fracture particles, are present in a paracrystalline arrangement throughout the entire
structure. There are a number of different types of cellular interactions

which involve direct physical contact between cells.

Today, let us consider a type of interaction that results in the formation of a low-resistance pathway. Since the late 1950's, it has been demonstrated that cells interact directly to form a low resistance pathway. Further, this pathway has been associated with an electrical interaction between cells that can be characterized as an electrical synapse. This type of interaction occurs without the synaptic delay that one observes in a chemical synapse.

In the next decade, several laboratories confirmed the presence of this type of interaction between a variety of cells, both excitable and non-excitable. To date, this phenomenon is found in virtually all metazoan cells, with the exception of circulating blood cells and mature skeletal muscle cells.

A couple of years ago, Ray Reeves, Alan Steinbach, and I were intrigued by the presence of two separate pieces of information, both of which concerned the phenomenon that required direct physical contact between cells. One type of interaction was the low resistance pathway or electrical synapse, and this is generally referred to as ionic coupling between cells. The other phenomenon, described by Subak-Sharp and his collaborators in Scotland, was the phenomenon of 'metabolic cooperation between cells.' In this particular phenomenon, there is an interaction between two different cell types (one a genetic variant) in which there is a metabolic capacity conferred on the genetic variant or mutant.

So we selected three cell types with which we could task the following questions: (1) can ionic coupling and metabolic cooperation occur simultaneously; (2) if they do, what is the structural apparatus or pathway for this communication; and (3) are there cell types where one can demonstrate that the failure to interact ionically and metabolically is accompanied with the inability to form a particular kind of pathway? We used a normal cell from the DON hamster cell line, and with this DON cell, we had two genetic variants: The DA cell (a variant of the DON cell) and the A9 cell, which is a variant of the mouse L cell line. The DON cell is completely normal with respect to its capacity to incorporate the exogenous purine ^{3}H hypoxanthine into its nucleic acids, whereas the DA and A9 cells fail to demonstrate this capacity. The DON cells have inosinic pyrophosphorylase activity which is not present in the DA and A9 cells. By co-cultivation of those cell types, one can determine that in one case the mutant cell is capable of incorporating exogenous labels into its nucleic acid only after a physical interaction with the wild type cell (CON).

One can observe the metabolic cooperation between these cell types by using radioactive labels and co-cultivating the wild type and mutant cells. In this first case, the DON cell is quite heavily labeled and the DA cell appears to incorporate the label after interaction with the normal cell. On the other hand, the A9 cell, which is the mutant derived from the L cell line, never appears capable of incorporation of the label even after direct physical contact with the wild type cell. In this case we had two variants, one which could cooperate metabolically and one which could not. By impaling the cells with microelectrodes and recording the voltage attenuation by passing a current from one cell to the adjacent interacting cell, we could record the degree of ionic coupling. In the first case (DON : DA), there is a voltage deflection which indicates a presence of a low-resistance pathway between the two cell types. If you examine the DON:A9 interactions, you fail to observe a low resistance pathway or ionic coupling. Therefore, the data indicated that when there is failure of ionic coupling, there is also failure of metabolic cooperation.

We examined the different populations of the cells carefully to determine the structural pathway. We found, with respect to a structural pathway, that there was a close membrane apposition between cells in those populations which communicated both ionically and metabolically. This particular apposition is characterized by two plasma membranes of adjacent cells (which are around 70 to 80 Å) which are separated by a so-called space or "gap" which is approximately 20-40 Å. This structure was appropriately named the gap junction by Revel and Karnovsky in 1967. Within the space or gap, there is a periodic series of dense connections or particles. With freeze-fracturing, the intramembrane content of the gap junction is characterized by a lattice of closely packed intramembranous particles. These particles within the junctional plaque appear to have a homogenous size (about 85 Å in diameter). (Figure 30.)

From this study, we were able to conclude that when metabolic and ionic coupling are present, there is a specific site of membrane specialization which is present to provide a "channel" mechanism for these functions. From its physiological nature, these so-called "channels" would have hydrophilic characteristics. From these experiments, it was apparent that it was important to look at gap junctions biochemically. Some efforts in this respect had already been initiated by a number of labs, and with remarkable success by Goodenough and Stoekenius (1972). We selected rat liver as the source of material for isolating gap junctions. We used the procedure which was first introduced by Neville for isolating an enriched plasma membrane fraction. From this material, we then exploited the detergent resistance of the junction which was described by Goodenough and Stoeckenius (1972) for isolating an enriched fraction of gap junctions from rat liver.

The enriched gap junctional preparation contains gap junctions which are not detectably altered by the isolation procedure. There is some contamination in the preparations which appear as amorphous material and small fragments of non-junctional plasma membrane. Figure 31.

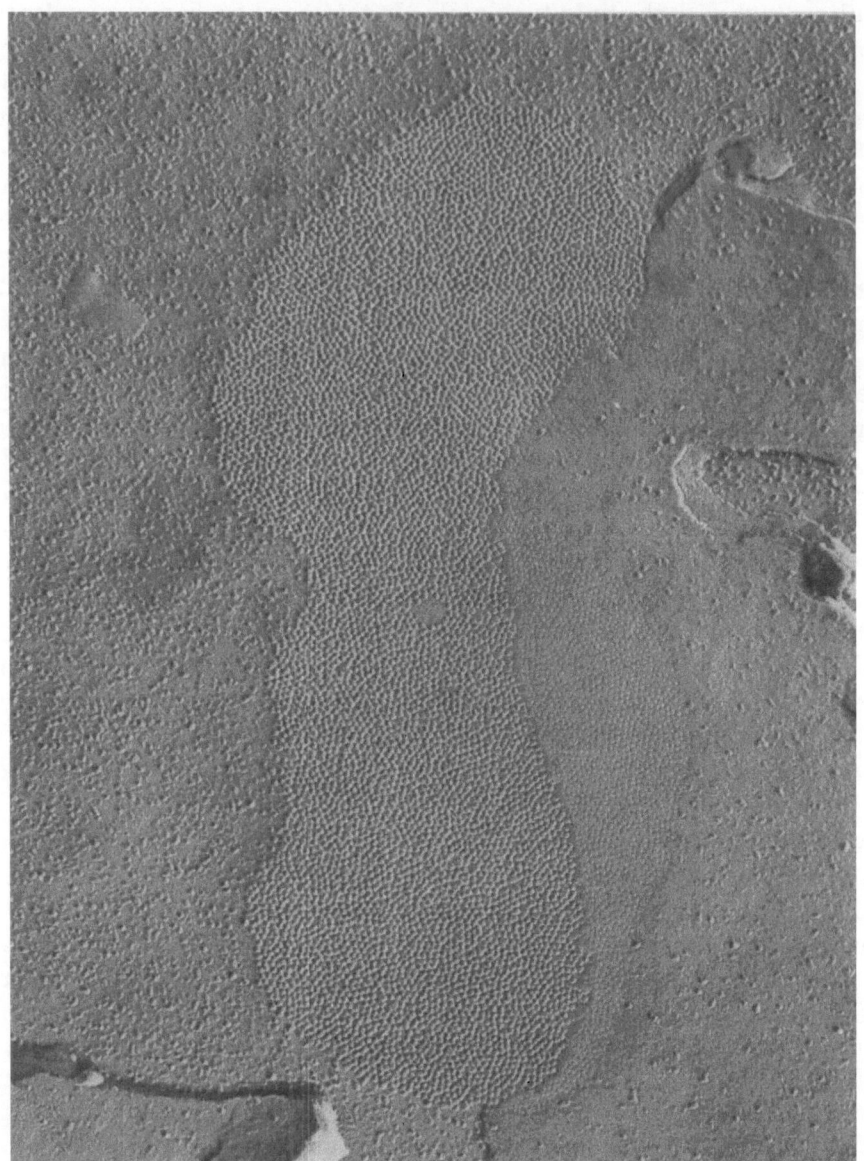

Figure 30. Freeze-fracture appearance of a gap junction between intact hepatocytes. The gap junction occurs as a plaque-like arrangement of polygonally-packed intramembranous particles and pits. The gap junctional intramembranous particles are homogeneous in size (about 80 A in diameter) which distinguishes them from the heterogeneous population of particles which are found in the adjacent non-junctional plasma membrane regions. X 98,400.

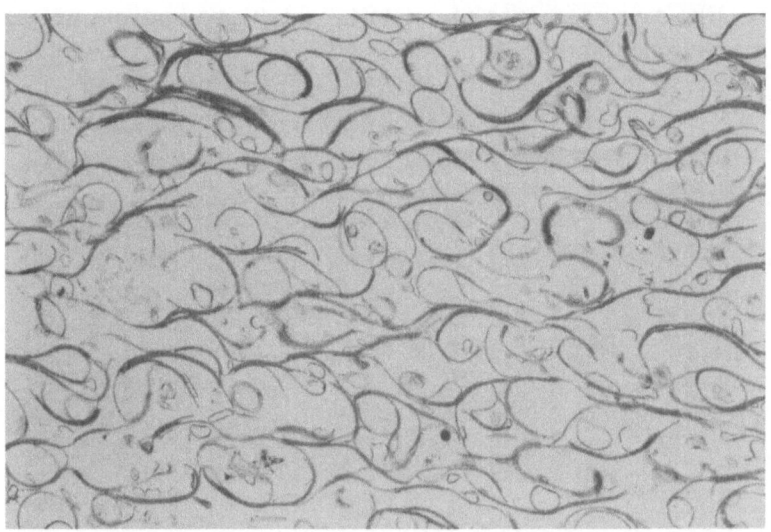

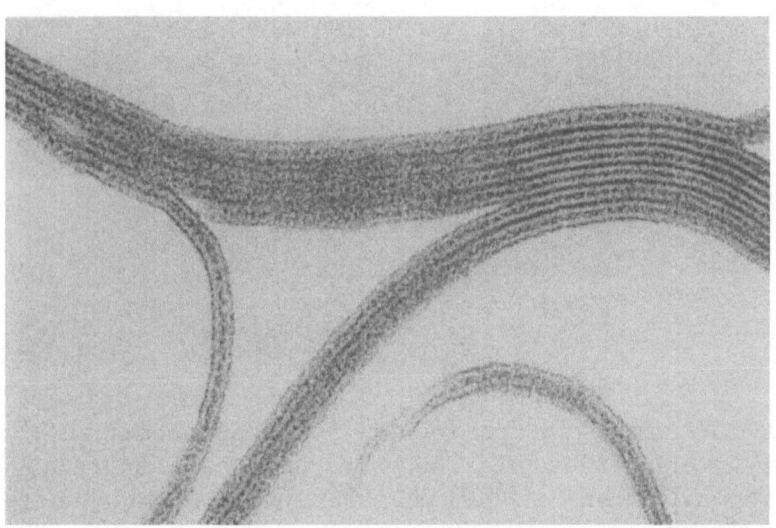

Figure 31. (Top). Thin-section appearance of the enriched gap junctional fraction from rat liver. At low-magnification, the preparation appears to consist primarily of single gap junctions and thick band-like structures. X 30,780.

(Bottom). High magnification appearance of the enriched gap junctional fraction. The band-like appearance in figure is due to the apposition of several gap junctions, producing a myelin-like image. The morphological contamination in these preparations usually consists of small, non-junctional membrane fragments and non-membranous, amorphous material. Note that the 20-40 A gap is still present in the isolated gap junctions. X 240,000.

      With negative staining, the gap junctional sheet is comprised of a lattice-like arrangements of subunits which are approximately 80 to 85 Å in diameter.　Figure 32.

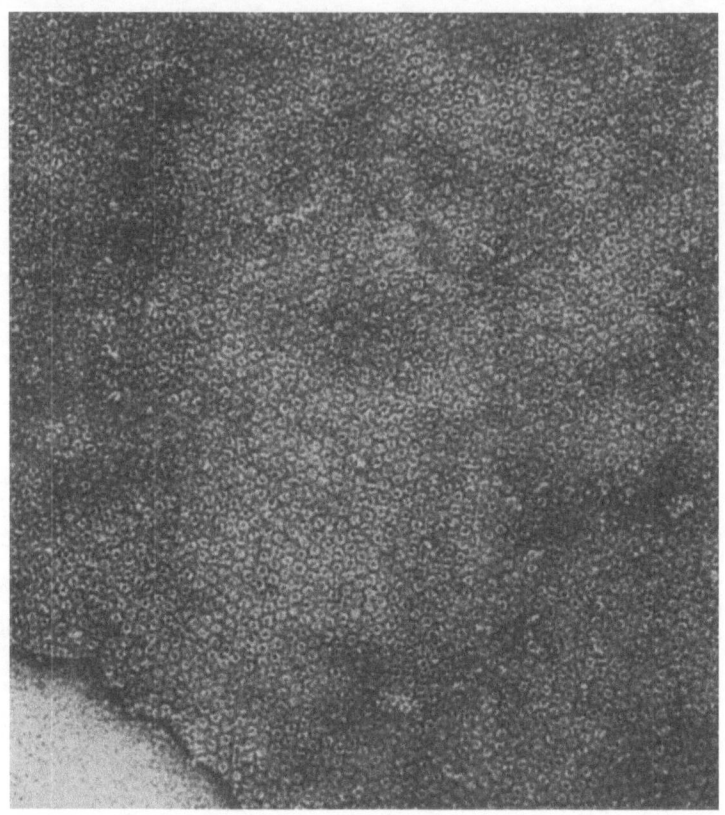

Figure 32.　Negative stain appearance of liver gap junction after treatment with uranyl acetate at low pH.　The entire junctional membrane is comprised of a paracrystalline arrangement of subunits 80-90 A in diameter.　A small, 20-25 A, electron dense dot occupies the central region of the 80-90 A particles in this preparation.　X 216,000.

      It is very difficult to determine yield because we cannot assay an endogenous activity associated with the gap junctional preparation.　However, we can determine the protein content of the enriched plasma membrane fraction and thereby, by comparing the protein content of the enriched gap junctional fraction, we can make an estimate of how much work protein is obtained.　A typical preparation of 400 grams of intact liver yields 760 mg of enriched plasma membrane; and from that, we obtain about 360 mg of enriched gap junctional protein.

After solubilization in SDS and reduction with DTT, there are two prominent components in coomassie blue stained 5.6 percent Fairbanks gels. The apparent m a s s of these two components is approximately 20,000 daltons and 10,000 daltons. They are extremely low molecular weight components, and we have been concerned with the possibility of proteolysis being present during our isolation preparation. Figure 33:

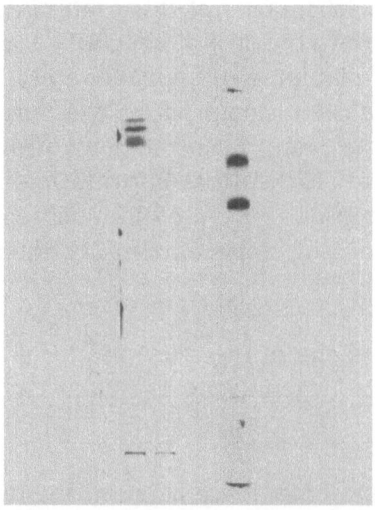

Figure 33: Electrophoretic separation of the rat liver gap junctional proteins. (Migration from top to bottom.) Left: Separation in a 5.6% polyacrylamide disc-gel system after solubilization in 2% sodium dodecyl sulfate and 40 mM dithiothreitol. Right: Separation in an exponential slabgel acrylamide system (7.5% - 15%).

To date we have no evidence to suggest that in fact we have proteolytic fragments in these two bands. At present, we have been unable to detect the presence of carbohydrates with either PAS staining or ninhydrin detection or amino sugars.

To determine whether or not these polypeptides are similar, we sliced the gel bands and analyzed their amino acid contents. With the cooperation of Jack Saari, we determined the amino acid content of the intact junction material, and we calculated a so-called polarity index as a means of expressing the polar amino acid content of the various proteins. From the polarity index, the content of the total fraction indicates a polarity of around 45 percent and this is a "normal" polar content for a variety of membrane proteins.

However, if we examine the content of the 20,000 dalton component and compare it with the 10,000 dalton components, we find the 20,000 dalton components is significantly different from the lower band, particularly in terms of its apolar amino acid content.

Recently, we have tried to consider the possible heterogeneity within each of these different polypeptide components, and we have examined this material in other gel systems. In the Neville continuous system where there is a gradient of acrylamide from seven percent to 15 percent, one can separate four components. At the present time we can consider the junction as a paracrystalline lattice. From the physiology, we know that there must be a site within this lattice for a channel, for hydrophilic continuity. In negative stain images of this material, particularly with uranyl acetate which at pH 4, there is a centrally stained region within the 85 Å subunit which has a diameter of approximately 20 to 25 Å. This dot is the tentative location of the channel within the lattice.

In the future we plan to continue our efforts to characterize the structure and translate the biochemical information back into the lattice, and also to use the information to develop immunological probes which can be used to study the process of biogenesis of communication between cells in culture.

EDELMAN: I think this is one of the most beautiful examples of a special and important structure in a membrane analyzed right down to molecular and structural levels.

BRANTON: What is the stoichiometry between the two proteins?

EDELMAN: Three moles of the 10,000 molecular weight to one mole of the 20,000 molecular weight.

KENNEL: A question that I wanted to direct to both of you and Dr. McConnell was in the analysis of these membranes, is there any evidence of lysophosphatide?

MC CONNELL: Does that have any effect on fluidity of the mix compound?

GILULA: I have no evidence of lysophosphatides present in the gap junction.

KENNEL: These structures look very much like chloroplast lamelli and mixed micelles formed from mixtures of lecithin and lysolecithin.

GILULA: In those lipid systems in which you can see rearrangements in the lipids, you don't get the intramembranous integrity that we see with freeze fracturing.

KENNEL: I absolutely agree that the pore structure is probably due to the proteins. I just wonder what the possible role of lipid is in this structure.

MC CONNELL: I think in all the systems I have discussed a lysolipid of any kind would be simply a detergent and would increase the fluidity. What percentage of the area looked at is made up of plaques?

GILULA: All of the structures which we examine occur in plaques, and in rat liver. Some morphometric data suggests that this structure can comprise as much as four to six percent of the total plasma membrane surface area.

EDELMAN: Do they move?

GILULA: They do move, and we have observed, particularly during tissue dissociation, that these structures have a remarkable mobility; however, the mobility that we observe is domain mobility, so that the entire structure moves as a lattice.

UHR: What is the effect of pharmacologic agents on the domain?

GILULA: We do not know the effects of pharmacological agents on the mobility of these junctional domains. The intriguing thing immunologically is that you can generate these structures between cells in different organisms as well as cells from different organs within the same species, so that apparently these particular products are encoded as a universal type product. Lowenstein and his collaborators have some information now that there is a hybrid possessing a certain chromosomal defect associated with failure of this cellular communication.

EDELMAN: We'll have a break and then a general discussion.

COFFEE BREAK

EDELMAN: Having some kind of basis or consistent picture for some of these cell types, we can now begin to discuss aspects of function and genetic control. I think everyone would agree that the three categories I have put on the board, cell division, cell movement, and cell-cell recognition, have in a time-honored sense been related one way or another to functions of the cell surface. In having the other speakers present their material, you might consider these categories even though we are far from a molecular understanding of these phenomena. I will call first on Bruce Cunningham, then each speaker will have a discussion, and finally we hope to have a general discussion. We are not going to let this run much later than five of one--and that will already be late.

CUNNINGHAM : We are interested in the mechanisms by which molecules acting at the cell surface can stimulate mitosis in lymphocytes. One of the biggest problems is our lack of understanding regarding the nature and the molecular interactions of most of the receptors on the cell surface. In the case of lymphocytes, there is one molecule we do understand fairly well, and that is immunoglobulin. All of the structural data indicate that the immunoglobulin molecule is organized into a series of compact domains with a true symmetry axis through the center of Fc portion of the molecule. Each of the domains (V_H, V_L, C_L, C_H1, C_H2, C_H3) is paired: V_H with V_L, C_H1 with C_L, C_H2 with C_H2, and C_H3 with C_H3. The main structural features of these domains are that they contain about 100-120 amino acid residues, and they contain a single intrachain disulfide bond forming a loop of about 60 residues in the polypeptide chain.

Yesterday you heard about a molecule, beta-2 microglobulin which is related to Ig and which appears to be closely associated with or to be a polypeptide chain of the histocompatibility antigen. The next slide shows the complete amino acid sequence of beta-2 microglobulin which we have determined. (Figure 34)

1   10   20
Ile-Gln-Arg-Thr-Pro-Lys-Ile-Gln-Val-Tyr-Ser-Arg-His-Pro-Ala-Glu-Asn-Gly-Lys-Ser-

21   30   40
Asn-Phe-Leu-Asn-[Cys]-Tyr-Val-Ser-Gly-Phe-His-Pro-Ser-Asp-Ile-Glu-Val-Asp-Leu-Leu-

41   50   60
Lys-Asp-Gly-Glu-Arg-Ile-Glu-Lys-Val-Glu-His-Ser-Asp-Leu-Ser-Phe-Ser-Lys-Asp-Trp-

61   70   80
Ser-Phe-Tyr-Leu-Leu-Tyr-Ser-Tyr-Thr-Glu-Phe-Thr-Pro-Thr-Glu-Lys-Asp-Glu-Tyr-Ala-

81   90   100
[Cys]-Arg-Val-Asn-His-Val-Thr-Leu-Ser-Gln-Pro-Lys-Ile-Val-Lys-Trp-Asp-Arg-Asp-Met

Figure 34: Amino acid sequence of human β_2-microglobulin.

The protein contains exactly 100 amino acid residues, and it contains two half-cystine residues. The size of the protein, therefore, is about the same as the Ig domains. The size of the loop resulting from the disulfide bond between the two half cystine residues is also about the same size as the intrachain loops in the Ig domain. If you compare the amino acid sequence of this molecule (Figure 35) with those of the various domains of IgG, they are very homologous:

```
                                          1                              10
β₂-MICROGLOBULIN                        ILE GLN ARG THR PRO LYS ILE GLN VAL TYR SER
EU C_L    (RESIDUES 109-214)            THR VAL ALA ALA PRO  -   -  SER VAL PHE ILE
EU C_H1   (RESIDUES 119-220)            SER THR LYS GLY PRO  -   -  SER VAL PHE PRO
EU C_H2   (RESIDUES 234-341)            LEU LEU GLY GLY PRO  -   -  SER VAL PHE LEU
EU C_H3   (RESIDUES 342-446)            GLN PRO ARG GLU PRO  -   -  GLN VAL TYR THR
```

```
                                                          20
ARG HIS PRO ALA  -  GLU  -   -   -   -  ASN GLY LYS SER ASN PHE LEU ASN CYS TYR VAL
PHE PRO PRO SER ASP GLU GLN  -   -  LEU LYS SER GLY THR ALA SER VAL VAL CYS LEU LEU
LEU ALA PRO SER SER LYS SER  -   -  THR SER GLY GLY THR ALA ALA LEU GLY CYS LEU VAL
PHE PRO PRO LYS PRO LYS ASP THR LEU MET ILE SER ARG THR PRO GLU VAL THR CYS VAL VAL
LEU PRO PRO SER ARG GLU GLU  -   -  MET THR LYS ASN GLN VAL SER LEU THR CYS LEU VAL
```

```
              30                                          40
SER GLY PHE HIS PRO SER ASP ILE GLU VAL  -   -  ASP LEU LEU LYS ASP GLY GLU ARG ILE
ASN ASN PHE TYR PRO ARG GLU ALA LYS VAL  -   -  GLN TRP LYS VAL ASP ASN  -  ALA LEU
LYS ASP TYR PHE PRO GLU PRO VAL THR VAL  -   -  SER TRP ASN SER  -  GLY  -  ALA LEU
VAL ASP VAL SER HIS GLU ASP PRO GLN VAL LYS PHE ASN TRP TYR VAL ASP GLY  -   -  VAL
LYS GLY PHE TYR PRO SER ASP ILE ALA VAL  -   -  GLU TRP GLU SER ASN ASP  -   -  GLY
```

```
              50                                          60
GLU LYS VAL  -  GLU HIS SER ASP LEU SER PHE SER LYS ASN  -  TRP SER PHE TYR LEU LEU
GLN SER GLY ASN SER GLN GLU SER VAL THR GLU GLN ASP SER LYS ASP SER THR TYR SER LEU
THR SER GLY  -  VAL HIS THR PHE PRO ALA VAL LEU GLN SER  -  SER GLY LEU TYR SER LEU
GLN VAL HIS ASN ALA LYS THR LYS PRO ARG GLU GLN GLN TYR  -  ASP SER THR TYR ARG VAL
GLU PRO GLU ASN TYR LYS THR THR PRO PRO VAL LEU ASP SER  -  ASP GLY SER PHE PHE LEU
```

```
              70                                          80
TYR SER TYR  -  THR GLU PHE THR PRO THR  -  GLU LYS  -  ASP GLU TYR ALA CYS ARG VAL
SER SER THR LEU THR LEU SER LYS ALA ASP TYR GLU LYS HIS LYS VAL TYR ALA CYS GLU VAL
SER SER VAL VAL THR VAL PRO SER SER SER LEU GLY THR GLN  -  THR TYR ILE CYS ASN VAL
VAL SER VAL LEU THR VAL LEU HIS GLN ASN TRP LEU ASP GLY LYS GLU TYR LYS CYS LYS VAL
TYR SER LYS LEU THR VAL ASP LYS SER ARG TRP GLN GLN GLY ASN VAL PHE SER CYS SER VAL
```

```
                          90                              100
ASN HIS VAL THR LEU SER GLN PRO  -   -   -  LYS ILE VAL  -  LYS TRP ASP ARG ASP MET
THR HIS GLN GLY LEU SER SER PRO VAL THR  -  LYS SER PHE  -   -  ASN ARG GLY GLU CYS
ASN HIS LYS PRO SER ASN THR LYS VAL  -  ASP LYS ARG VAL  -   -  GLU PRO LYS SER CYS
SER ASN LYS ALA LEU PRO ALA PRO ILE  -  GLU LYS THR ILE SER LYS ALA LYS GLY
MET HIS GLU ALA LEU HIS ASN HIS TYR THR GLN LYS SER LEU SER LEU SER PRO GLY
```

Figure 35: Comparison of the amino acid sequence of β₂-micro-globulin with the amino acid sequence of the homology regions of the immuno-globulin G, Eu.

For example, there are 11 positions where the Ig domains are all homologous to each other. Beta-2 microglobulin has identical residues in 10 of the 11 positions.

The histocompatibility system has a very interesting genetic locus. In the mouse, the H-2 system has two major loci, the K locus and the D locus. Between these loci are a number of other loci including the immune response genes, and the gene for a serum protein, Ss. On the basis of the structure of immunoglobulin, it was proposed that immunoglobulin evolved by duplication of a precursor gene large enough to specify a molecule about the size of beta-2 microglobulin. Based on these findings and the genetics of the immune system, Edelman and Gally proposed a hypothesis for the arrangement of Ig genes into a series of gene clusters, called translocons. They proposed that there are three such translocons, one for kappa chains, one for lambda chains, and one for heavy chains. In addition, they suggested there might be a fourth translocon that would be more primitive than the other three and would represent immune response genes. They suggested, in fact, that all of these genes might arise from a primitive precursor of the histocompatibility genes. The findings that beta-2 microglobulin is the appropriate size and is homologous in sequence to Ig, and that this protein is associated with the histocompatibility antigen suggests that this hypothesis may be correct.

It is, therefore, particularly important to find out whether the gene for beta-2 microglobulin is located on the histocompatibility gene complex. In addition, we would like to know in detail how beta-2 microglobulin is associated with the histocompatibility antigen. Beta-2 microglobulin generally is found as a monomer; in other words, as a free, single domain. As I indicated before, all the Ig domain are paired. One key question is whether beta-2 microglobulin has a homologous partner on the heavy chain of the histocompatibility antigens, or if the heavy chain simply has a binding site for beta-2 microglobulin. The latter case would be like the complement system and be less interesting. In any event, it is important to emphasize that there is a primitive Ig-like molecule on all cells. This finding may have particular significance in considering the function and the evolution of all types of all receptors.

I would like to come back now to cell stimulation in the lymphoid system. One would prefer to stimulate lymphoid cells with specific antigens that would interact with Ig receptors. Detailed analysis, however, is extremely difficult because very few lymphoid cells (0.1%) are stimulated by any specific antigen. On the other hand, a group of plant proteins termed lectins appear to bypass the requirements for specificity and can stimulate a large percentage of the cells. We have been using predominantly one lectin, Concanavalin A (ConA). We have considered a number of mechanisms for how ConA might act. These are outlined on the next slide (Figure 36):

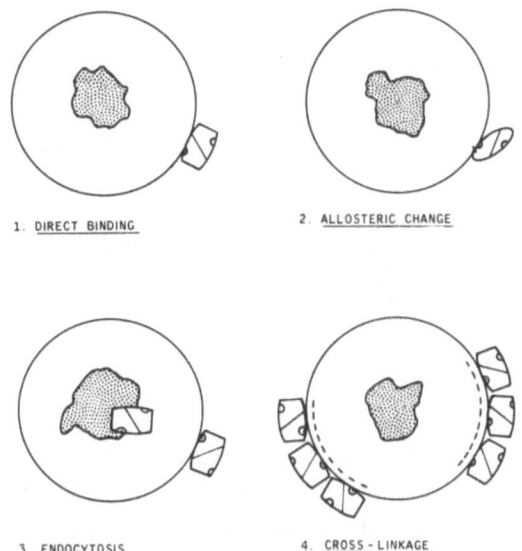

Figure 36: Possible mechanisms for the mitogenic stimulation by ConA.

ConA binds to cells by binding to carbohydrates on glycoproteins or glycolipids on the cell surface. Binding appears necessary for activity because the mitogenic activity can be inhibited by simple sugars such as mannose or glucose. It is difficult to believe that the direct binding to a carbohydrate moiety would provide changes sufficient to trigger a response; however, it is possible that binding might result in an allosteric change in ConA, and expose a second binding site which would be responsible for activation. Alternatively, ConA might bind at the cell surface and following binding, be taken into the cell where it then exerts its action, for example, in the nucleus. Finally, ConA on the cell surface might cross-link various receptors and stimulate activation in any of a number of ways. Although no mechanism can be given at this time, there are a number of facts that we do know about ConA stimulation. First, stimulation does not appear to require external factors such as components of serum. Secondly, it does not appear that the cell-cell contact is required.

From this information, we conclude that the cell has all the necessary apparatus to respond and the major initiator is the interaction of ConA with cell receptors. It, therefore, seems particularly important to consider the kinetics of the response and the populations of responding cells.

The next slide (Figure 37) illustrates an experiment in which parallel cultures of spleen cells were stimulated with ConA, and at various times α-methyl manoside was added to cultures to inhibit ConA binding, and remove ConA from the cells:

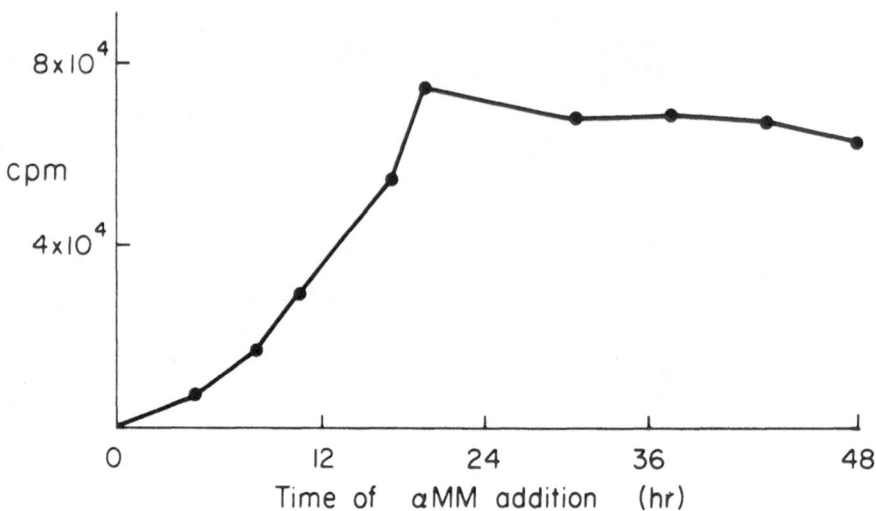

Figure 37.   Inhibition by alpha-methyl-D-mannoside of the incorporation of ^{3}H-thymidine into lymphocytes stimulated with Con A.

If Con A is removed about twenty hours after stimulation, the maximum response is observed. Before 20 hours, the response increases with time. Control experiments suggest that this increase is not due to an increase in irreversibly bound Con A. In addition, the addition of methyl mannoside does not appear to change the rate of DNA synthesis. The question is whether the increased response with time is the result of all cells being stimulated to a greater and greater extent, or whether more and more cells are being recruited.

Gary Gunther and John Wang have carried out an elegant study in which they repeated this experiment, but in addition to measuring the incorporation of labeled thymidine into DNA, they counted the number of blast cells and the number of labeled blasts at each point. They also determined the number of grains in each labeled blast. The results are illustrated in the next slide (Figure 38):

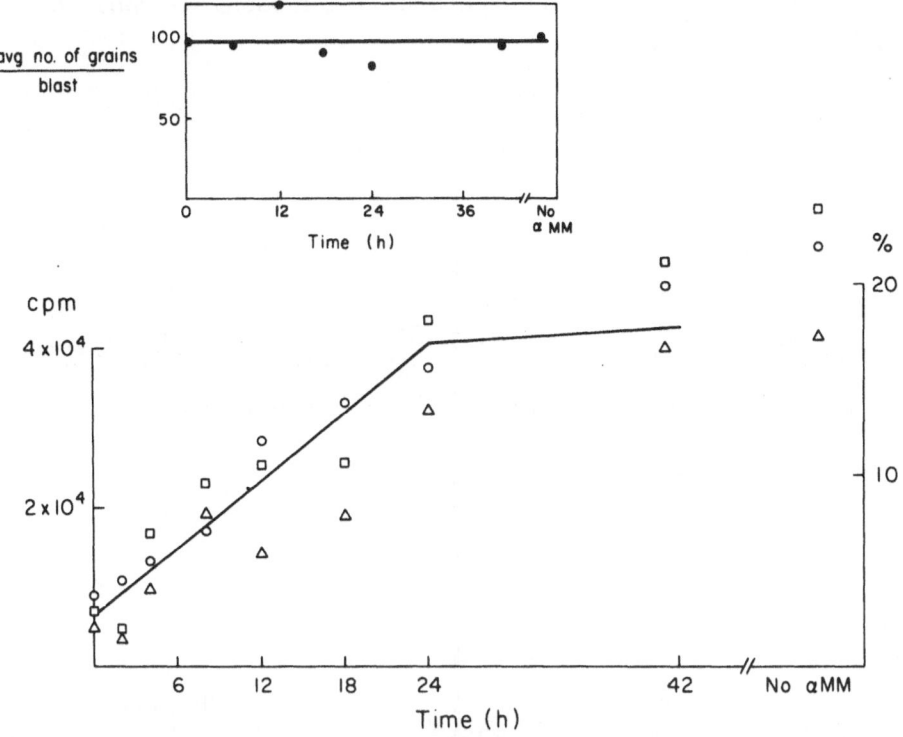

Figure 38. Comparison of total ^{3}H-thymidine incorporated (o, left ordinate), % blasts (□, right ordinate) and % labelled blasts (△, right ordinate) at various times after stimulation of lymphocytes with Con A. <u>Insert:</u> The number of grains per labelled blast.

The open circle represents incorporation of tritiated thymidine into DNA, the squares denote the percent blast cells and the triangles indicate the percent labeled blasts. All follow essentially the same line. In addition, the insert shows that the average number of grains in each of the labeled blasts is not changed over the period of time.

From these data we make a number of conclusions. First, it appears that the rate of DNA synthesis is independent of the length of exposure to Con A. Secondly, increasing numbers of cells seem to be committed on longer and longer exposures to Con A. Finally, cellular commitment appears to be an all or none phenomenon, in that once a cell is stimulated, it goes on to synthesize as much DNA as it is going to make, regardless of whether Con A is present or not. We have considered two models to account for these findings, and these models are shown on the next slide (Figure 39):

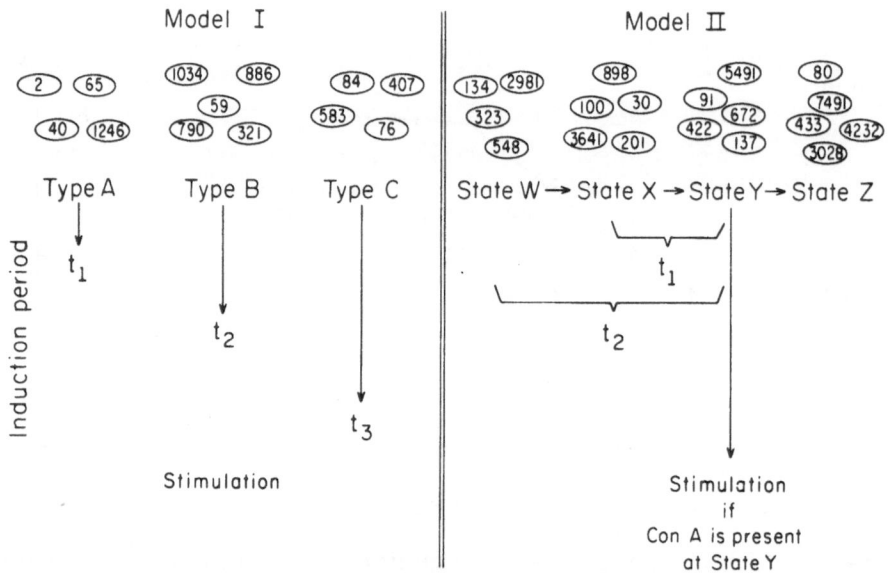

Figure 39. Two possible alternative models to account for the kinetics of the stimulation of lymphocytes by Con A.

The first model suggests that the lymphocyte population is extremely heterogenous with multiple types of cells. Each different cell type would require a longer exposure to Con A in order to be stimulated. Alternatively, the lymphocyte population may be reasonably homogenous, but cells could exist in any of a number of temporal states. In this model, stimulation by Con A could take place in only one of these states, e.g., state Y. Cells that were not in state Y, would cycle to state Y at different times. Therefore, over a period of time, you would recruit more and more cells as they entered state Y.

To test this hypothesis, cells were exposed to two separate pulses (6 hours each) with Con A and the incorporation of tritiated thymidine into DNA was compared with that obtained with single six-hour pulses at different times. The first model predicts that the second pulse should not result in increased stimulation, provided, of course, that there is no memory. The second model predicts that the second pulse would result in twice as much incorporation as in a single six-hour pulse. The results of this experiment are illustrated in the next slide (Figure 40):

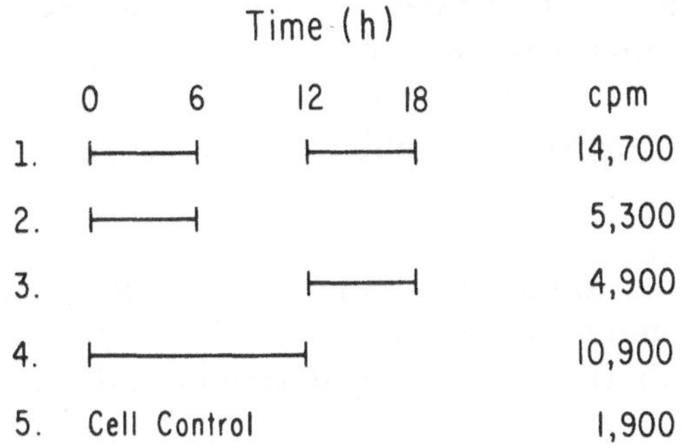

Time (h)

	0	6	12	18	cpm
1.	⊢——⊣		⊢——⊣		14,700
2.	⊢——⊣				5,300
3.			⊢——⊣		4,900
4.	⊢—————⊣				10,900
5.	Cell Control				1,900

Figure 40: Incorporation of ^{3}H-thymidine into mouse splenocytes after exposure to ConA for: 1. two six-hour pulses separated by six hours without ConA; 2. and 3. one six-hour pulse, and 4. one continuous twelve-hour pulse.

The cells exposed to two six-hour pulses of ConA gave more than twice as many counts as those exposed to single pulses. From these data we form Model II which suggests that the resting lymphocyte population is distributed among a number of various states. This hypothesis is in accord with results obtained with other cells. Temin has studied the stimulation of chicken fibroblasts with serum, and he concluded there is a similar type cycle in these cells. I should emphasize we are talking about a distribution throughout G_I. We are not talking about a distribution throughout S, G-2, and M, which are presumed constant for all of the cells. Smith and Martin have suggested a general hypothesis in which cells exist in two states, a determinate (A) state in which the cells proceed through the normal cycle at a fixed rate, and an indeterminate (B) state in which the cells are randomly distributed throughout the G_I phase of the cycle.

Cells are recruited from the indeterminate state on a probability basis. The data we have obtained provide support for this hypothesis and suggest that the ability of lymphocytes to respond to ConA is related to some indeterminate state in G_I.

EDELMAN: Now this is open for discussion. I think one of the implications of this kind of study is that it is rather difficult to do biochemical studies involving 20 hours of exposure of a whole population to ConA and then up all these heterogeneous things to see what you can dig out. One of the questions I had is that if this turns out to be true in general, what does it mean?   Why do we ever want to have a cell population distributed this way?

TOMKINS: There is another interpretation of Smith and Martin's model. They conclude that there is an "A" phase and a "B" phase, B represents the portions of the cycle which are essentially determined, A is not necessarily differentiated, but going from A to B is lowering a probability function.

CUNNINGHAM: I think your formulation was correct, I didn't phrase it properly.

EDELMAN: Gordon, I think your point is well taken; that is the best way to describe it. But the problem remains: what is the underlying biochemistry, how can you distribute this population of cells, and what do the triggers do?

HAKAMORI: With regards to the susceptibility of Con A induced transformation at G_1 phase, which was just presented by Dr. Cunningham, I would like to mention our recent studies (with Dr. Gahmberg) on surface exposed glycolipid and glycoprotein during the cell cycle of NIL cells. Synchronized NIL cells (hamster fibroblasts) exhibit maximum label on galactoprotein and various glycolipids at G_1 phase. The labels in globoside, which has the terminal N-acetylgalactosamine residue, and in ceramide trihexoside, which has a terminal β-galactosyl residue, increase several-fold at G_1 phase, although the chemical quantities of these glycolipids were unchanged. Studies by other investigators showed incorporation of radioactive precursors into glycolipids increased greatly at G_1 phase.   These studies clearly showed that the organization and synthesis of these membrane receptors greatly changed during the cell cycle.

TOMKINS: They would say the A state is totally undifferentiated.

EDELMAN: That is a kind of phenomenological description.   I think Tomkins tends to confirm Smith and Martin's hypothesis.

LERNER: I would like to take a minute to describe the slime mold system. A perturbation is made on a population of cells that are alike genetically (a clone) when you take away their food. One cell sits down and calls forward the rest.   Since we are dealing with a clone, there is no reason why one or another cell should sit down; yet, only one cell sits down at a time and, therefore, I think we are stuck with a probability phenomenon.

EDELMAN: It is probably normally distributed. What I think is most interesting is the evolutionary implications of this distribution. It is because cells send up scouts every once in a while to see what they should do vis-a-vis the population? Is it that cells don't like to have the whole critical switch point in one place to be disrupted by some virus, and therefore distribute themselves in a series of protective states? Perhaps we are too far in advance of the data.

BACH: I wanted to come back to the beta-2-microglobulin which was mentioned at the beginning of this talk and your notion that this may be related to the Ir genes. Two questions come up both of which related directly to what you said: one, if it is the Ir product, you would expect to find allelic forms of this, because there are alleles of Ir which control immune response. The second, does anybody have any information whether allelic forms of beta-2 have been found?

EDELMAN: I can take care of this and save some time. We don't have any. Does anybody else have polymorphic forms of beta-2? (No response) Well, I don't suppose that anybody would suggest that beta-2 is an Ir gene product in view of its homogeneity. Let me call on Tomkins, because he comes right in here.

TOMKINS: I have been interested for a long time in how signals from outside influence intracellular behavior. One system we have been working with is illustrated in Figure 41.

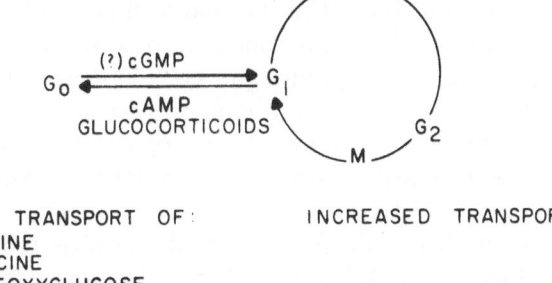

Fig. 41. Role of glucocorticoids and cyclic nucleotides in regulation of cell cycle events.

The cells were derived at the Salk Institute by Horibata and Harris. They are called S-49 lymphoma cells and are TL-and theta-positive Balb-/C mouse cells. Their most interesting property, from our point of view, is their sensitivity to a variety of hormones. I have been mainly

interested in the adrenal steroid hormones which a r e cycle-specific in -
hibitors of growth. They block cells in a state I have illustrated here as
G_0. This could equally well be formulated as a point in G_1 or in the "A"
state of Smith and Martin. A number of biochemical events take place as
a result of this block: transport processes are decreased, macromolecular synthesis is decreased, and there is an increase in intracellular proteolysis. Another event which takes place is that the cells die. We have
been using this property of these cells in order to isolate mutants (which
I shall discuss presently). If we remove the glucocorticoids, cells which
are n o t dead re-enter the cycle. There is a concomitant change in t h e
"pleiotypic" biochemical parameters, similar to what happens when PHA
stimulates a population of immunocytes or when a n y mitogenic stimulant
is given to a population of cells sensitive to it.

We have been studying two stimuli w h i c h affect growth, the
glucocorticoids and the cyclic nucleotides, or agents which stimulate their
modulation. A reasonable formulation of the mechanism of steroid action
is t h e following: the steroids penetrate the cell membrane and i n t e r a c t
with cytoplasmic protein receptor molecules. T h e r e are about 10^4 re -
ceptors per target cell. The r e c e p t o r s are allosteric proteins w h i c h
alternate between two forms, "active" and "inactive". An active steroid
stabilizes the active conformation of the receptor.

Following this interaction, there is a second reaction, which
we call "activation" of the receptor-steroid complex. Activation a l l o w s
the molecule to associate with an acceptor site in the nucleus. As a result
of this interaction, new messenger molecules appear in the cytoplasm.
Hormone antagonists interact with the inactive conformation of the receptors, and the subsequent events do not take place. These "antihormones"
can thereby inhibit the action of an agonist. Suboptimal agonists produce
suboptimal effects even at very high concentration. They have affinity both
for the active and inactive forms of the receptor molecules, so they establish a constant ratio between active and inactive forms which determine
the suboptimal level of activity.

The activated receptor-steroid complex leaves the cytoplasm
and goes into the nucleus. When the steroid is removed, the sequence is
reversed and receptor returns to the cytoplasm.

Figure 42 shows the necessity for activation. R e c e p t o r
steroid complex disappears from the cytoplasm and appears very slowly
in the nucleus at zero degrees. If we raise the temperature to 37°C, then
there is a rapid movement of the receptor out of the cytoplasm, and into
the nucleus. If you wait for several hours and then shift the temperature,
then the translocation takes place very rapidly at that time. This temperature-dependent step appears to be activation, since the receptor-steroid
complex forms readily at zero degrees.

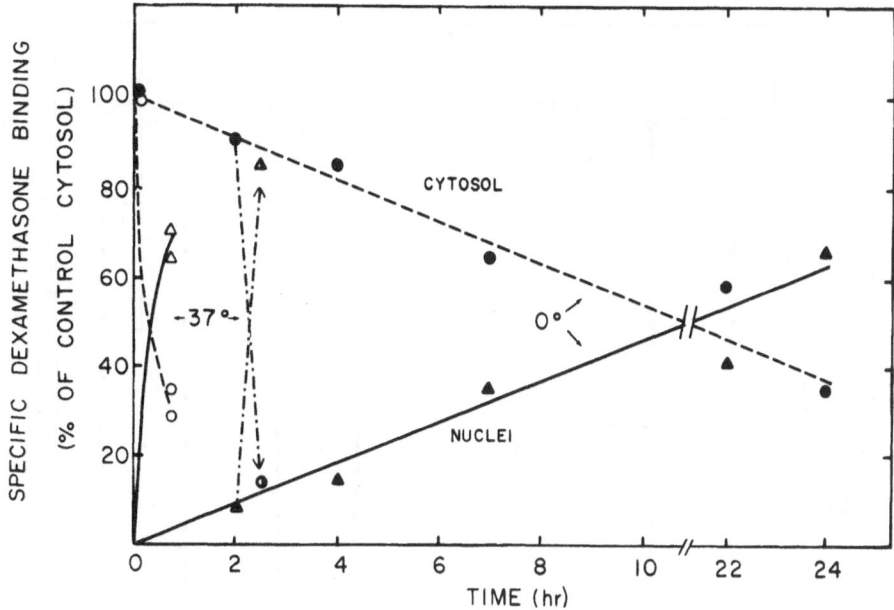

Fig. 42. Dexamethasone binding as a function of time at 0⁰ (▲–▲ nuclear activity,●—● cytosol activity), at 37⁰ (△–△ nuclear activity,○—○ cytosol activity), and after rapid shift of temperature (▲—▲ nuclear activity and ◑—◑ cytosol activity.

The next Figure (43) illustrates a point about partial agonists. The height of the bars represents the amount of receptor present in the cytoplasm in the absence of steroid. In the p r e s e n c e of an active inducer, most of the receptor leaves the cytoplasm and receptor-steroid complex appears in the nucleus. Progesterone, which a n t a g o n i z e s the action of inducers, associates with the receptor, but does not cause t h e receptor to go to the nucleus. Deoxycorticosterone, a partial a g o n i s t, causes only partial nuclear localization. The main question in this area is: what are the nuclear sites? Receptor-steroid complexes bind to chromatin. So we hope to use receptors as probes of chromatin organization as well as for figuring out how steroids work.

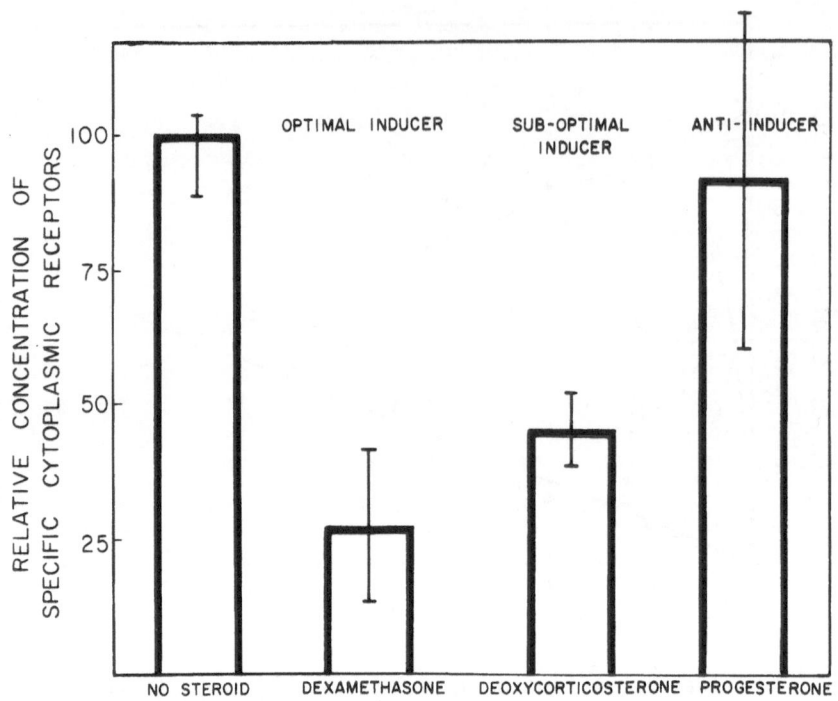

Fig. 43.   Relative concentration of specific cytosol receptors in the presence of various steroids.

            So we have been trying to superimpose a genetic approach
on chemical work.   S-49 cells are killed by the glucocorticoids, and the
killing represents the binding curve of the receptors.   The nice part of
this is that there is an occasional resistant colony.   Carol Sibley was
able to isolate many resistant clones.   Fluctuation analysis showed that
steroid resistance arises as a stochastic process independent of the presence of the selective agents.   The frequency of the transition is 10^{-6}/
cell/generation.   It is increased significantly by different kinds of mutagens.   The range of steroid-resistant mutants selected corresponds well
to our presumptions about the mechanism of hormone action.   The majority are mutants in which the steroids receptor capacity is missing. These
are called r^-.   There are also mutants in which the   receptor-steroid
complex can be formed, but in which the complex does not associate with
the nucleus.   We call these "nt-," or nuclear transfer minus.   Finally,
there is a class in which all the steps take place, but in which the physiology of the cell is not affected.   We originally called these d-, (for
"deathless").

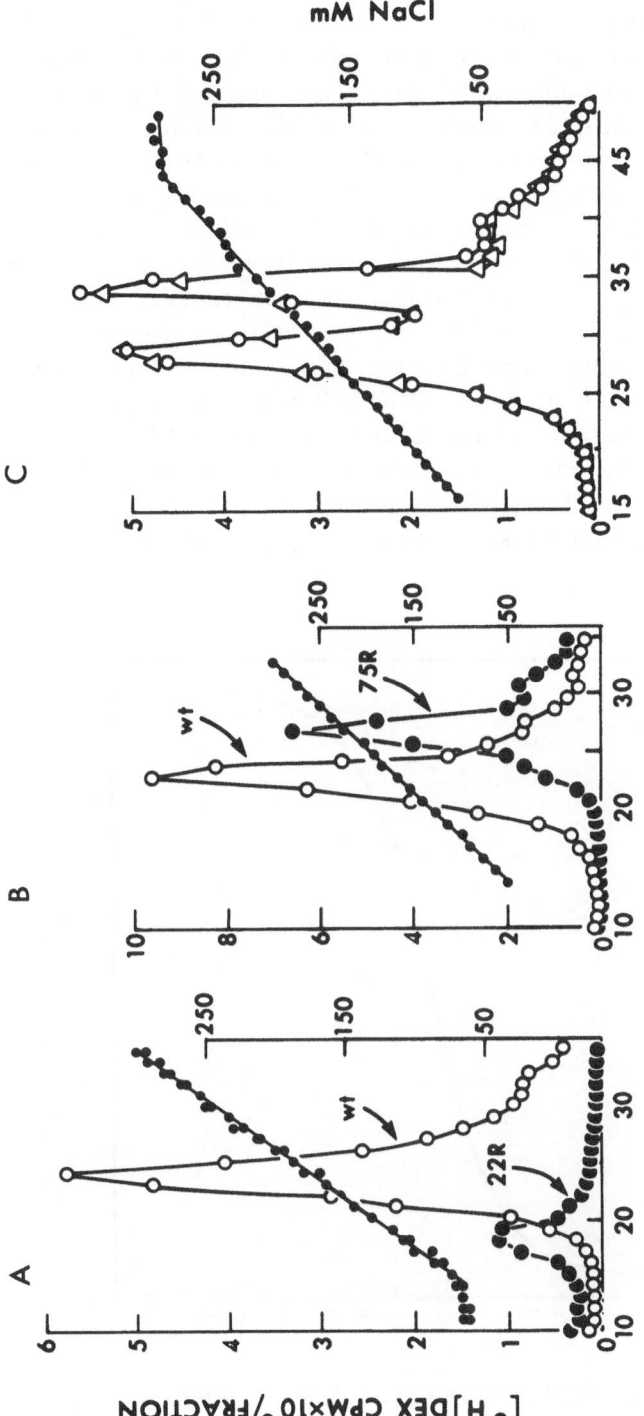

FRACTION NUMBER

Fig. 44. Elution of DNA-cellulose bound receptors with 50-300mM NaCl gradients. In each case, extracts in TEGNO 5 were loaded pairwise onto separate DNA-cellulose columns. After being rinsed free of unbound material, bound receptors in the two columns were eluted with NaCl gradient drawn from a single mixing chamber. Fractions approximately 250 µl were collected directly into counting vials; the NaCl gradient of each fraction. A: o—o, wt; (wild type)●—● 22R. Identical results are obtained using 55R. B: the o—o, wt; ●—● 75R. Identical results are obtained using 13R. C: o—o wt + 75R, incubated at 20°C prior to mixing. △—△ wt + 75R incubated at 20° (after mixing identical results are obtained using 55R.

The mutants we have studied thus far have all turned out to be mutants in the steroid receptor molecule. We have used these altered molecules to try to determine the nature of the nuclear site for receptor binding. A great many experiments which preceded this work has led us to believe that the receptor-steroid complexes are DNA-binding proteins. Receptor-glucocorticoid complexes bind to DNA, they prefer double-stranded DNA, they don't bind to RNA, but we have not found any sequence specificity for the DNA binding. However, reports from other laboratories have suggested that complexes may bind to proteins. So we have been using the mutants to get at this question. One type of mutant receptor-steroid complex does not accumulate in the nucleus in intact cells, does not bind to the nucleus in cell-free binding experiments, and as shown on the next slide (Figure 44), there is weak binding to DNA cellulose. Therefore, receptors which don't go into the nucleus don't bind very well to DNA.

We have also managed to find mutant receptors which bind excessively well to the nucleus. The DNA binding of these receptors is likewise enhanced. So the conclusion of this work is that receptor-steroid

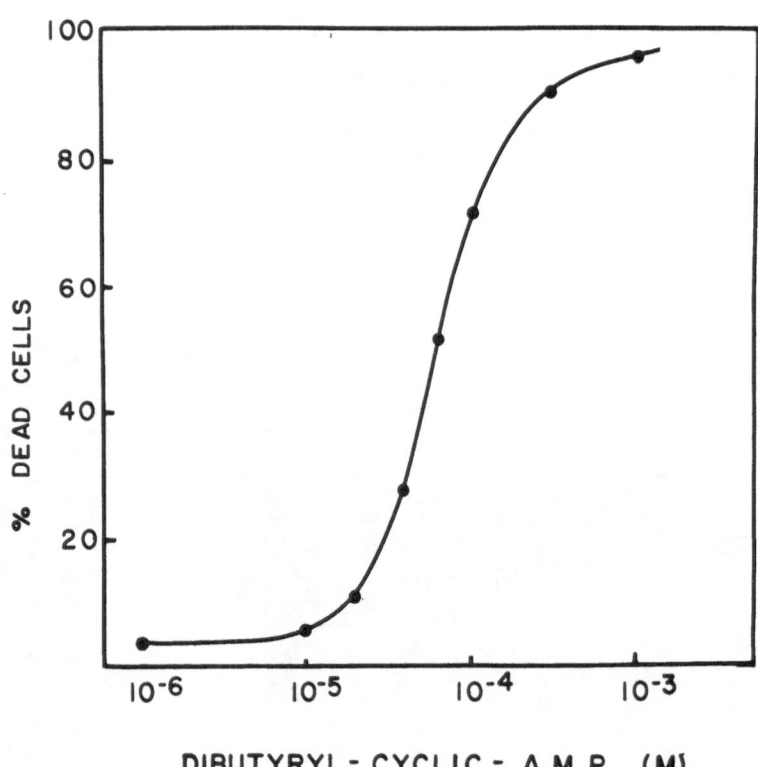

Fig. 45. Effect of dibutryl-cyclic AMP on S49 lymphoma cells.

complexes bind to nuclear DNA which is their site of action. We don't know whether they recognized specific sequences or structural peculiarities in the DNA, nor do we know what happens when the DNA binding takes place.

I would also like to discuss briefly the mechanism of killing. This seems to be a physiological action of the glucocorticoids on the immune system. Cell hybridization experiments have suggested that the steroids induce lethal molecules and we investigated whether these might be related to cyclic AMP. We found first that the glucocorticoids stimulated c y c l i c AMP accumulation slightly (perhaps by repressing the phosphodiesterase). Second, we asked whether cyclic AMP itself was lethal to the cell. Figure 45 shows this to be the case. Figure 46 shows the effect of dibutryl-C-AMP.

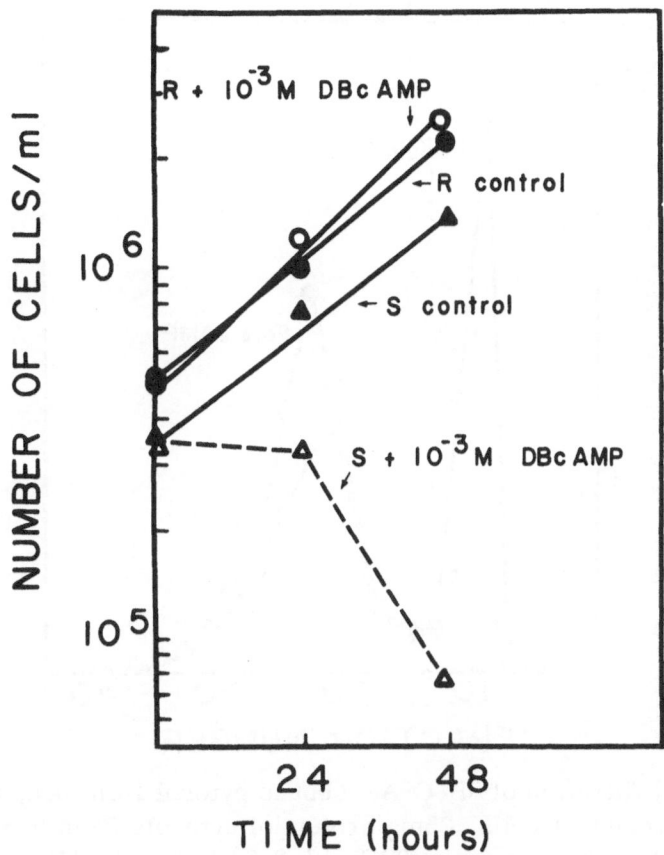

Fig. 46: Effect of but_2C-AMP on the growth of sensitive and resistant lymphoma cells. Cells sensitive to but_2-C-AMP were grown in the absence (▲—▲) or presence (△—△) of 1mM but_2-C-AMP and 0. 2mM theophylline. Cells resistant to but_2-C-AMP were g r o w n in the absence (●—●) or presence (○—○) of the same reagents.

Dibutyryl cyclic AMP, like the gluco corticoids is cell cycle-specific in that cells are inhibited in G_1 (or G_0) prior to killing. The fact that dibutyryl cyclic AMP kills the cells raised the possiblity that we could isolate resistant mutants. We were especially interested to determine whether the steroids function via the cyclic nucleotides. Philip Coffino has isolated mutants resistant to an AMP cycle and has calculated the mutation rate to be 10-7 cell/generation, independent of the selective agent and increased by mutagens. The next slide shows the growth of S-49 cells. The doubling time is 16 to 18 hours. Including AMP, causes the death of sensitive but not resistant cells, which continue to grow. Figure 46 shows cyclic AMP binding activity in wild-type and mutants clones.

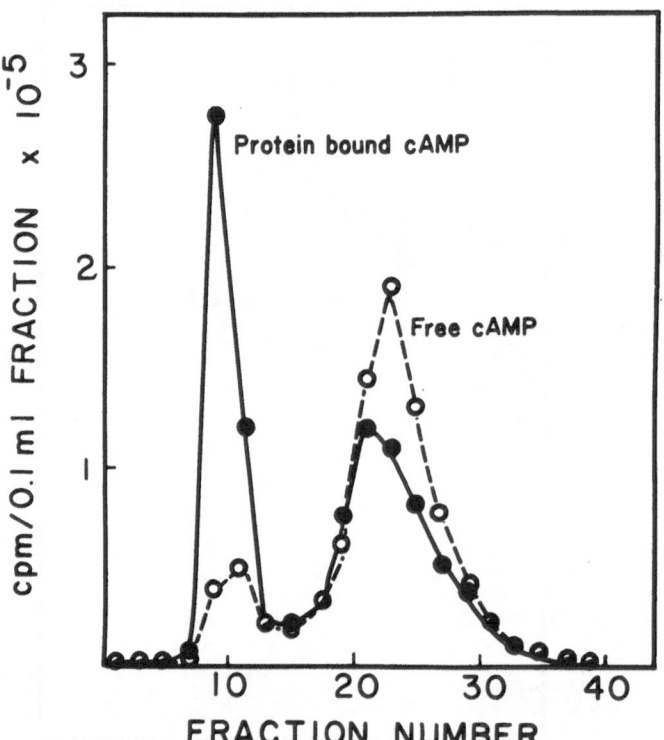

Fig. 47. Gel filtration of 3H-C-AMP bound cytosol from but_2-C-AMP-sensitive and resistant cells. 25mg protein of cytosols from both lines were incubated for 60 minutes at 0^0C with 0.5mM 3H-C-AMP in the presence of 5mM theophylline. (Sensitive cells ●—●, Resistant cells ○—○.)

Everyone of the latter so far analyzed has been different both in cyclic AMP binding protein and its associated protein kinase. There are many physiological consequences of this work which I don't have time to discuss, but I can say that we have been using these techniques as a means of trying to isolate mutants defective in membrane hormone receptors.

EDELMAN: Just one question related to these corticoid receptors. Are they connected in any way to the membrane?

TOMKINS: No.

LERNER: Have you looked at the reversion frequency of the mutants?

TOMKINS: We have just developed a method for selecting revertants.

EDELMAN: Since the time is short, I am going to call on Dr. Moscona in the hopes that we can round it all up now.

MOSCONA: From what has been said so far at this meeting, it is obvious that the cell surface is a very versatile organelle both in terms of its composition, structure, and function. I want to bring into this discussion still another aspect of the cell surface, which is of great important for everyone. It has to do with mechanisms which make it possible for single cells to associate with each other, to interact, and to construct multicellular tissues and organs. I am referring, of course, to selective adhesion and recognition of embryonic cells.

Much was said at this meeting about immunological recognition, and I should perhaps point out how it differs from the concept of embryonic cell recognition. Immunologic recognition is displayed only by lymphoid cells, and its function is to protect the genetic integrity of the organism from foreign intrusion. Embryonic cell recognition, on the other hand, is a property present in all tissue cells, and its function is to integrate the cells into morphogenetic structures. In simple terms, it is a mechanism which enables the different kinds of cells in the embryo to identify each other by contact, to aggregate selectively and become organized into tissue-forming groupings. When this mechanism fails to function properly in the embryo, due to genetic or environmental causes, morphogenesis is abnormal, and this may lead to congenital malformation: in the adult organism, loss of cell recognition is often said to accompany neoplastic invasiveness and metastasis.

Embryonic cell recognition and selective cell adhesion can be studied experimentally by the procedures of in vitro reaggregation of embryonic cell suspension. (For a review, see Moscona, 1973.) Some of you probably know that, when an embryonic tissue is treated with trypsin the enzyme cleaves intercellular bonds, the cells round-off and can be dispersed into a suspension. However, dispersed embryonic cells do not "like" to stay separated, and under suitable conditions of tissue culture they reaggregate rather rapidly into multicellular systems. In 24 hours or less, the single-cell suspension is converted into a group of multicellular aggregates. Within the aggregates, the cells become organized: they sort out, and reconstruct their characteristic tissue pattern.

If the suspension contains cells from two different tissues, the different kinds of cells sort out and segregate either in separate aggregates, or in different regions of composite aggregates. For example, in aggregates which are made to contain a mixture of embryonic liver cells and cartilage cells, the different cells segregate and form distinct masses of liver and cartilage tissue. Similarly, in composite aggregates of retina and liver cells, the different cells sort out into tissue-specific groupings. Evidently, cells from different tissues recognize and distinguish between homologs and heterologs, and accordingly they establish histospecific contacts conducive to interactions, morphogenesis, and differentiation. We proposed, as a working hypothesis, that embryonic cell recognition and selective cell adhesion are mediated by interactions of specific cell-membrane components which function as recognition-markers and specific cell ligands: we suggested that cell-ligands from different tissue might differ in terms of their qualitative characteristics, amounts on the cell surface, and topographical distribution (Moscona 1962, 1968, 1973). The implication from this was that, cells with complementary ligands would display positive recognition and would become linked into a tissue: while absence of ligand complementarity, or absence of ligands, would result in transient cell adhesion, non-specific adhesion, or no adhesion.

These assumptions led to two predictions. First, if cells from different tissues possess qualitatively different ligands, one might expect cells from different tissues to show different surface antigenicities. Second, it might be possible to isolate from embryonic cells materials with tissue-specific ligand properties; the addition of such materials to suspension of aggregating homologous cells would be expected to enhance the reaggregation of the cells, and this effect should be tissue-specific and should result in the formation of much larger cell aggregates than in controls. These predictions were tested with positive results. By immunizing rabbits with suspensions of live cells from various embryonic tissues, we were able to prepare tissue-specific antisera directed selectively against the cell surface. For example, an antiserum prepared against a suspension of live embryonic retina cells, and thoroughly absorbed with non-retina cells, agglutinated only retinal cell suspension; when coupled with fluorescein, this antiserum stained only the surface of retina cells. Similarly, antiserum prepared against liver cell suspension reacted, following absorption with non-liver cells, selectively with surfaces of embryonic liver cells. In fact, several such tissue-specific, cell-surface directed antisera were prepared (Goldschneider and Moscona, 1972).

These findings demonstrated the presence of tissue-specific determinants on embryonic cell surfaces; since their immunological disparities corresponded to the recognition-specificities of these cells, we proceeded to a more direct test of the cell-ligand concept, i.e., isolation from cells of materials with tissue-specific cell-ligand activity. Procedures for the preparation of such materials were developed first for embryonic retina cells. We took advantage of the fact that embryonic cells maintained in primary monolayer cultures continuously resynthesize surface materials, and shed some of them into the medium (Lilien and Moscona, 1967; Hausman and Moscona, 1973). Fractionation of these products lead to the isolation of a fraction which, when added to aggregating suspensions of embryonic retina cells, enhanced cell reaggregation, resulting in the formation of considerably larger aggregates than in the controls. This effect was tissue-specific, and dose-dependent. Subsequently, other tissue-specific cell-aggregating factors were prepared. For example, a cell aggregating factor prepared from embryonic cerebrum cells enhanced only the reaggregation of embryonic cerebrum cells, but not of cells from other brain regions, or from the retina, or other tissues.

Further purification of the retina-specific cell ligand preparation by Sephadex gel chromatography and DEAE fractionation led to the isolation of a glycoprotein fraction, which on SDS-polyacrylamide gels bands in a single zone, in a region corresponding approximately to 50,000 M.W. The retina-specific cell-aggregating activity is associated with this glycoprotein (McClay and Moscona, 1974). The amino acid and sugar composition of the purified preparation was examined by Dr. Hausman. The carbohydrate part represents somewhat less than 10 percent of the total and consists of glucosamine, mannose, galactose, and sialic acid. Desialation of this material did not destroy its specific activity, nor was it destroyed by short treatment with periodate. Tryptic digestion destroyed ligand activity. Antiserum prepared against this purified retina cell-ligand material reacted specifically with retina cell surfaces, not with cells from other tissues. We have presented elsewhere further evidence which suggests to us that this retina-specific cell-aggregating glycoprotein is functionally representative, and perhaps identical, to the native cell-ligands of the cell surface.

It is not yet known if the 50,000 M.W. particle is the actual ligand, or a sub-unit of a larger complex; nor do we know, at present, how it functions to cross-link cells. These are some of the problems being studied.

It has been suggested by several investigators that aggregation of neural retina cells is mediated by cell-surface galactosyl-transferase and that it represents the tissue-specific cell-ligand. We therefore examined if the activity of our retina cell-aggregating preparation was due to the presence of galactosyl-transferase activity or receptor activity for this enzyme. We found that our purified material contained neither. Therefore, the cell-linking activity of our preparation evidently involves other kinds of glycoprotein associations with the cell surface. However, this finding does not exclude the possible role of galactosyl-transferase in the biosynthesis or turnover of retina cell-ligands, or in the anchoring of cell-ligands to the cell surface; this problem is now being investigated.

Further progress with this problem will require preparation and biochemical characterization of specific cell-ligand materials from several tissues, analysis of their association with cell surfaces, and of their exact mode of action. Of particular interest should be information on the distribution patterns of cell-ligands on cell surfaces, since it is quite conceivable that their specific effects depend not only on chemical disparities, but also on differences in their topographies on the cell surface.

EDELMAN: Now we are open for discussion.

GILULA: Do you have any idea whether the amount of specific factor increases or decreases during aggregation? In other words, do you find a difference in the level of aggregation factor that is produced by small aggregates versus large aggregates?

MOSCONA: Normal re-aggregation of trypsin-dissociated tissue-cells requires regeneration of the ligand-material on the cell surface. In the absence of such regeneration the cells do not re-aggregate histotypically. We have not attempted to measure with precision the regeneration of this material during normal cell aggregation, because the small amounts involved and the technical difficulties in this kind of measurements.

TOMKINS: There was a report several months ago, I don't remember the author, but I am sure you do, where the Sperry experiment could be done in vitro. Have you looked at the effect of your factor on that?

MOSCONA: We have not. These experiments were done by Roth and his associates. They studied specific associations between retinal cells and optic tectum. Their results agree well with our general scheme of the mechanism of cell recognition.

YAHARA: I think definitions of recognizing molecule and recognized molecule are relative. But if your factor is a recognizing molecule, what molecule is recognized by your factor?

EDELMAN: Aaron, the question is, even if the distinction between a recognizer and the molecule recognized is understood to be a relative one, what does the recognizer recognize? Do I have that right?

MOSCONA: It is a bit premature to speculate about details that are now being studied. Theoretical models have the danger of being taken too seriously. The experiments to answer your question can be done and the results should be forthcoming.

KARNOVSKY: As cells mature into adulthood, does the factor persist, or does it disappear, and are there then structural complexes such as cell functions involved?

MOSCONA: This is very important. It must be recognized that cross-linking factors are the first step in establishing tissue associations themselves. Once this step has taken place, other types of junctional complexes such as specialization of the cell surface, take over resulting in the formation of the intercellular relationships which provide a differentiated tissue.

LERNER: I have been wondering through the whole meeting, is there any evidence for lectins in eukaryotic membranes such as occur in lower forms?

EDELMAN: Lectins? You are saying the plasma membrane has lectins?

LERNER: Again, in slime molds, there is clearly a lectin that may function as receptors during differentiation.

EDELMAN: I don't think people think of lectins as being product of the plasma membrane. The horseshoe crab has a molecule in its blood that we worked on in our lab some time back. It is a specific hemagglutinin and it has a specificity for sialic acid. We don't know where this thing comes from. It is sort of interesting, but I know of no other case.

POLLARD: I just had one question to direct to Dr. Moscona. Since this protein is about 50,000 in weight, it gives me an idea what its size might be. I don't know how close together these embryonic cells approach and how many contact points they have, but if they don't approach too closely or if they contact only in a few places, then several of these proteins must get together to form chains. How many proteins would be necessary to bring the cell membranes together in order to effect specific contact?

MOSCONA: The distance between the cells is approximately 200 Å . I don't know whether the 50,000 M.W. units represent monomers of larger complexes which are the ligands, of whether each of these in itself function as a ligand. Let me point out that the approach itself is not a simple matter. We often think of cells as little globules, spheres, which make contacts at the defined points. With respect to tissue cells, this is not the case because the surface of cells is extremely complex, extremely active with a large number of contact points between cells. It is not clear whether there is a difference between the properties of the contact point and the properties of the rest of the membrane, so I am trying to say that any answer to this question would be oversimplification.

EDELMAN: Before we sum up, I should like to make a statement that I hope will provoke Aaron's response. I don't think there is any prima facie evidence that his molecule is a ligand. It could be a hormone, or it could be stimulating a whole set of other ligands which come out of cells. I see nothing so far in the evidence that this is acting as a multivalent ligand for the cells. I don't think you would claim that, would you, at this state? For example, it could induce the formation of another molecule which is a ligand or alter the membrane of the cell.

MOSCONA: That is the point.

RAFF: I would like to ask you a question regarding the biological significance of this type of gross tissue type cell recognition. Recent observations in the development of insects, such as drosophila, have shown that very early in development there is a progressive compartmentalization of the cells that are destined to form a particular structure--such as the wing. So that from very early on, before there is any wing, cells that will form anterior, posterior, central, dorsal, etc., parts of the wing are somehow determined, and there is absolutely no mixing of cells across the boundaries of a compartment when the wing finally forms. Now you find that all the cells of the retina, for example, recognize each other and form large aggregates, apparently overriding any finer type of recognition such as neuronal cells versus pigment cells or dorsal versus ventral. How do you reconcile your findings with these finer types of recognition which must exist?

MOSCONA: Your question relates to the physico-chemical aspects of this problem. To answer it would require, in my opinion, definitive structural-functional information about relationships between what we call the ligand material, and the cell surface. We are far from knowing this. I can only say that, at the present level of analysis, the evidence on hand--and it is, unavoidably, indirect evidence--strongly suggests that these ligand materials which we have isolated are derived from the cell surface, that they function at the cell surface, and that they may be the actual components of the cell-linking and cell-recognition mechanisms. Detailed information, at a physical level, about their mode of function will not be easy to obtain, certainly not without knowledge of their molecular structure.

EDELMAN: It seems to me, whatever the level of specificity of the investigation, that the mechanisms we have been talking about all day are related to the problem of recognition. Of course, we would all like to tie up to what we know about immunological recognition on the one hand, and what we know about histocompatibility and variability on the other, but so far no one has come up with either enough facts or enough ideas to penetrate through all of these fields. Nonetheless, I felt it important to call these phenomena to your attention.

Today I think we did make a connection with what was said yesterday. I think it is fairly clear that what we have to do now (and perhaps what the people who work on the red cell have to do) is to relate these proposed functional assemblies to some structural analysis. This would help to settle the question of how much specificity there is in the membrane assemblies of differentiated cells.

I am going to call this to a halt and thank the speakers and participants.

SESSION IV

BIOLOGICAL ACTIVITIES OF SOLUBILIZED CELL
SURFACE COMPONENTS

Major histocompatibility complex (SD, LD, Ia, Ir, Ss products) - Tumor specific transplantation antigens - Immunologic unresponsiveness - Autoimmune diseases - Hormone receptors - Hormone binding - Adenyl cyclase activation - Granulocyte colony stimulating activity - Chalones.

SESSION IV

BIOLOGICAL ACTIVITIES OF
SOLUBILIZED CELL SURFACE COMPONENTS

KAHAN: The biological activities of two types of surface membrane components have been extensively studied, namely, those which can serve as antigens to elicit an immune response and those functioning as hormone receptors. Among the former group, transplantation antigens, the substances which elicit the rejection of foreign grafts, serve as an example where solubilized derivatives have been shown to possess many of the same biological properties as the native cell surface material. Since the chemical properties of transplantation antigens have not been elucidated, it is not at present possible to identify these materials by structural or compositional characteristics. Furthermore, transplantation antigens can only be defined by their activity in the artificial situation of graft compatibility since their actual function in the economy of the cell surface membrane is unknown. There is no firm evidence to support the numerous roles which have been proposed: telegraphic messages of growth, cell surface virus receptors, transport proteins, pure cytostructural elements, or the immunologic role as primary markers of "self". Even more embarrassing than the deficiencies in our knowledge about the actual structure and function of these molecules, is our ignorance about the mechanism mediating the role of the surface structure in the recognition and destruction of the cell by immune elements. There has been no work of the type performed with hormones by Katzen, Cuatrecasas, and others (described below), showing specific binding of immunocompetent cell receptors to native or solubilized surface transplantation antigens. The second event in the recognitive process may be similar to the generation of adenyl cyclase, as Rodbell and Lin have studied following glucagon binding. Further, the translation of the message into the end biologic activity may proceed through the local action of mediators, such as the colony stimulating activity of Till and Price, which is released as a function of membrane events on the surface of leukocytes. This activity probably represents the final effector phase of a multi-step membrane process.

Unfortunately, the biologic assessment of these surface components relies upon an indirect, "artificial" event of transplantation. The in vitro analysis of soluble materials as allospecific transplantation antigens is based upon their capacity to induce or to interfere with the action of humoral or cellular immune components. Although the biological assessment of soluble tumor-specific transplantation antigens is even more primitive than that of allospecific markers, there is some evidence that the physiology of these substances may be important to the apparently protected status of neoplastic growths. As McKhann will point out, circulating tumor-specific transplantation antigens may control the immuno-regulation of the host's destructive response, and thereby favor the pro-duction of antibody protective to the graft. These findings are consistent with the hypothesized role of surface antigens as circulating telegraphic messages regulating growth, and/or reflecting the distinctive composition of "self".

After Fritz Bach has acquainted us with the molecular genetics of the transplantation system, referred to as the major histocompatibility complex, we will attempt in the discussion to pull together studies on the biological activities of solubilized cell surface membrane derivatives, both in vitro and in vivo, as outlined in Table 7.

Table 7
Identification of Transplantation Antigens

I. Chemical Analysis: distinctive structure, composition
II. Biologic Functional Activity
   A. Actual function unknown? receptor for growth, virus receptor, etc.
   B. Assessment of role in transplant immunity
      1. Capacity to receive signal-binding to immunocompetent cell
      2. Translation of message-such as by adenyl cyclase activation?
      3. Biologic effect-release of local? transmembrane, ?cytoplasmic mediator
III. Immunologic Analysis
   A. "In Vitro"
      1. Cellular immunity
         a. Proliferation
            1) Non-immune cells: mixed leukocyte culture (LD)
            2) Immune cells: antigen-specific lymphocyte mitogenesis (SD)
         b. Generation of effector cells by primary sensitization
         c. Performance of immunocompetent cells
            1) Aggregation to specific targets
            2) Killing of specific targets
            3) Release of immune mediators
      2. Inhibition of the action of Humoral Antibodies
         a. Cytotoxic Antibody (SD and Ia)
         b. MLC Blocking Factors (LD ? Ia)
   B. "In Vivo"
      1. Allograft immunity
         a. Induction specific antibody
         b. Second-set phenomenon
      2. Cutaneous delayed-type hypersensitivity
      3. Immunologic enhancement
      4. Immunologic unresponsiveness

BACH: The chromosonal region usually referred to as the Major Histocompatibility Complex, or MHC, represents on the order of 1/1000th. of the total genetic material. Yet, the general impression is generated that an inordinately large number of membrane, or surface structures on a cell such as the lymphocyte, may be determined by genes in the MHC.

My purpose in this talk is to broadly introduce the genetic, cellular and molecular biology of the MHC; I will attempt to do so by dividing the presentation into those three areas. Whereas pertinent information is available in several species, the mouse is the model in which some of the most critical information has been obtained. I will limit myself largely to studies done in the species, although where additional information is useful, I will attempt to mention that. Given the information we have, I think it fair to extrapolate between species to a rather great extent; the broad biological principles appear to be very similar in several different species.

Genetic Considerations : In mouse, the MHC is in the IXth linkage group on chromosome 17 and is referred to as the H-2 complex. A genetic map of the region is shown in figure 48. H-2 is divided into four regions, K, I, S and D (from left to right) each including a marker locus, H-2K, Ir-1A, Ss and H-2D respectively. In addition to these loci there are several other loci, some of which have been formally mapped as separate from the above four; others have been indentified by studying a number of different phenotypic traits.

Alleles of the H-2K and H-2D loci determine cell surface antigens which were originally defined serologically and can be referred to as the serologically-defined, or SD, antigens. They are presumed histocompatibility antigens as we will discuss later. Whereas in mouse only two MHC SD loci have been clearly defined, in man three such loci have been identified as noted in figure 49. The SD loci are highly polymorphic. In addition to the Ir-1A locus, the Ir-1B locus has been identified. Alleles of these loci control the immune response of the animal to a variety of antigens.

The Ss locus controls the quantitative levels of a serum protein. Mapping with this locus, and perhaps identical with it, is the control of the Slp antigen, the expression of which is under hormonal control, and quantitative levels of components of serum complement.

There are at least two loci of H-2 differences at which result in activation of cells in mixed leukocyte cultures (MLC). One of these has been mapped between Ss and H-2D, and thus between the two regions also; the second maps in the I region and has not been formally separated from the Ir-1A locus. These loci, since they were first detected by lymphocyte response in MLC are referred to as lymphocyte-defined, or LD, loci. Differences of the LD locus mapping in the I region generally lead to stronger MLC activation than those of the locus between Ss and H-2D; we thus refer to the former as the "strong" LD locus, or LD_S, and the latter as LD_W.

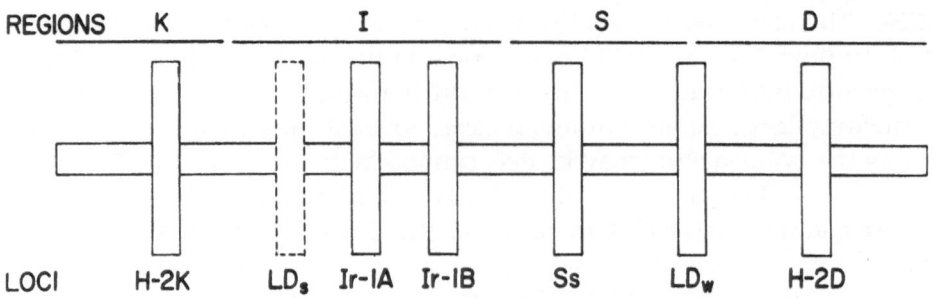

Fig. 48. The mouse H-2 complex is divided into four regions, K, I, S and D each designated by a marker locus, H-2K, Ir-1A, Ss and H-2D. The alleles of the H-2K and H-2D loci determine the classically serologically defined H-2 antigens. The Ir-1A and Ir-1B loci determine immune res-ponse of the animal and the Ss locus controls the quantitative levels of a serum protein. Differences of the two LD loci result in activation of lym-phocytes in mixed leukocyte culture and are associated with graft versus host reactions in vivo; the alleles of the LD-s locus are relatively stronger in this regard than alleles of the LD-w locus. The LD-s locus has not been formally separated from the Ir-1A locus. Differences of the LD-s locus are also associated with skin graft rejection without concomitant H-2K or H-2D disparity.

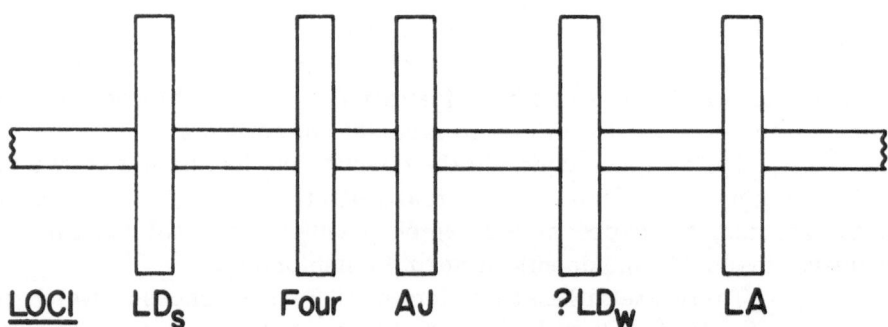

Fig. 49. Major Histocompatibility Complex of Man - HL-A:
A schematic representation of the major histocompatibility complex in man -- HL-A. There are three identified loci, LA, Four and AJ, alleles of which determine the serologically defined antigens. The locus differ-ences which lead to strong activation in MLC are mapped as LD-s; the LD-w locus, if it does exist, maps between the AJ and LA loci.

Eventually these loci will probably be designated by numbers.

Several groups have recently demonstrated that cell-surface antigens which are determined by genes in the I region can be detected serologically. These antigens behave differently from those referred to above as the SD antigens. The SD antigens of H-2K and H-2D are present on all lymphoid cells and essentially all other tissues examined; the antigens determined by I region genes, called Ia antigens, are present on only a minority of lymphocytes (primarily B lymphocytes); the exact distribution of the Ia antigens on other tissues has not been extensively investigated.

A number of other phenomena have been related to the MHC which may well function through, or be controlled by, cell membrane structures. These include, among others, Hybrid Resistance, which controls the ability of bone marrow transplants to "take" following transplantation, and susceptibility to certain leukemogenic viruses in mouse and disease susceptibility in humans (which may function in some instances in the same manner). To which extent these various phenomena are related to the MHC loci already mentioned is not known; I will not discuss them further since I do not believe that they represent problems which are readily related to cell membrane experimentation at the molecular level.

It is worthy of mention that, in addition to mouse studies, MHC SD loci have been identified in a very large number of species, that MHC LD loci have been identified (as separate from the SD loci) in humans, dogs and rhesus monkeys; that MHC Ir loci have been found in humans, rat, rhesus monkey and guinea pig among others. Once again, insofar as we understand the biology of these various loci, they are very similar in the various species.

Before leaving the genetic considerations, let me place the MHC in mouse in a broader context with respect to cell surface molecules. Fifteen recombinational units to the left of H-2 is the T locus, a genetic complex alleles of which are involved in developmental processes. It has been demonstrated that these genes also determine cell surface structures, at least on sperm. One unit to the right of H-2D is the TLa locus, also determining a cell-surface antigen. We should thus, perhaps, extend our concept of the genetic region which relates to cell membranes in this linkage group.

Cellular Considerations: In this section I want to review, again in broad outline, the biological processes to which the phenotypes of the various MHC genes discussed above can be related. It is essential that we remind ourselves that a given locus is mapped by a certain procedure measuring its phenotype; we can only critically relate the one phenotype in question to that locus. A different phenotypic trait can be related to the genetic region for which that locus serves as a marker; further for a variety of reasons we may assume that the locus is directly related to both phenotypes, however only molecular studies can serve to provide critical evidence in this regard.

The MHC SD antigens were defined serologically; they have been related to transplantation rejection and have, in fact, been referred to as the major H antigens. The Ir genes have been functionally mapped in the MHC; it has been hypothesized that they may determine the T lymphocyte receptor, although other hypotheses for their role in regulating the immune response have been suggested. Some of these other ideas are still consistent with the concept that the Ir genes determine a membrane component -- other than the primary T cell receptor.

The LD loci were defined when it was noted that cells of individuals differing for the MHC LD loci, but identical for the MHC SD loci, stimulate in MLC. It is presumed that the LD loci are represented on the cell surface although this has not been formally proven.

The Ia antigens are detected serologically; no known biological function has been related to them.

We can discuss the MHC LD, SD and Ir loci in their relationship to in vitro and in vivo models of allograft immunity and immune response. I elaborate on some of these models now since I believe that they will also serve to test for the biological functions of various isolated membrane components.

The MHC LD products are presumably responsible for activation of T lymphocytes in MLC; it is not clear to which extent the MHC SD antigens can, per se, result in activation of cells in MLC. The data in mouse is difficult to analyze since there is MLC activation when congenic mice differ by either the K or D regions; however it is not certain that this is caused by the SD antigens determined by the H-2K or H-2D loci. In man, differences for the SD loci with identity for LD_s results in very weak or zero MLC activation.

Cells activated in MLC become specifically cytotoxic to sodium chromate (^{51}Cr) labelled target cells taken from the same strain used as the stimulating (sensitizing) cell donor. This is referred to as the cell-mediated lympholysis (CML) assay, and is illustrated in Table 8. It is possible to demonstrate, however, that whereas the differences which are primarily responsible for activation of proliferating cells in the MLC are the LD differences, the primary target in CML is determined by the SD regions. This has perhaps been most elegantly demonstrated in a "three-cell" experiment done by Schendel in Madison; the results are shown in table 9. Strains AQR and B10.T(6R) differ by H-2 LD differences and are H-2 SD identical; AQR and B10.A differ for H-2K. Whereas there is an excellent proliferative response in the AQR-6R combination, there is no significant CML activity; i.e., the LD differences do not serve as a good target or do not lead to the generation of cytotoxic cells. In the H-2 SD different combination, there is some proliferative response and low CML activity (under some culture conditions this combination leads to no CML despite a good proliferative response). In a three-cell experiment in which AQR cells are stimulated by cells both of the 6R strain and by cells of B10.A, there is both a proliferative response and the development of cytotoxic cells -- the cytotoxic attack is, however, only against the SD different cells.

It is not clear whether the SD antigens themselves are capable of stimulating both the proliferative and the cytotoxic response, and whether it is the SD antigen itself which is the target in CML or whether the target is determined by another gene in the K region which is different from that of the H-2K locus. Nor are we helped very much by the availability of animals carrying mutations which affect the SD antigens or other loci in the MHC since we cannot be sure that these are point mutations which affect only a single locus.

We will return to a discussion of this apparent functional dichotomy of the MHC LD and SD components -- the LD differences leading to activation of a proliferative response in MLC, the SD acting as a target in CML later. I would emphasize that whereas the majority of published studies have, as I have just shown in Table 9, demonstrated that the LD differences alone in MLC do not lead to a cytotoxic response in CML, some preliminary findings by Peck and Schendel in Madison suggest that this may not always be the case, although the degree of killing on LD is not as great as that on SD.

The most critical data regarding the relative role of MHC LD and SD components in allograft phenomena has been obtained in mouse. It would appear that animals that differ for H-2 LD loci and are identical for SD, have strong proliferative graft versus host reactions; further that animals that are identical for the H-2 SD antigens can have fatal graft versus host reactions, and that the frequency of these reactions is related to the degree of MLC activation (although in these studies the MLC activation cannot be directly related to the MHC). Animals that differ for either the H-2 K or D regions, manifest much weaker GvH reactions between them.

Skin grafts are rejected when animals differ for the D or the K region alone, likewise there is rejection (in the few strains that have been tested) when animals differ for the LD_s locus. At least one combination which differs for LD_w shows long-term skin graft survival.

The conclusions from these studies, which are also consistent with the data which has been obtained in humans, is that in the non-sensitized recipient both LD and SD are important in graft survival, although some investigators believe that LD may be more important. For a proliferative GvH reaction, the LD differences are clearly more important -- as for MLC.

It should not be forgotten that since many of these correlations are being made with differences which are really "genetic region" differences and the loci in question were usually defined by phenomena other than in vivo allografting, that still other loci of the MHC may play an important role in

Table 8
Cell-Mediated Lympholysis

MLC Sensitization	Target Cells		
B10.BR $\pm$ B10	-17.2 ± 5.3*	-6.7 ± 6.6	19.1 ± 2.4
B10.BR $\pm$ B10.D$_m$	-17.7 ± 3.5	38.8 ± 4.6	-5.2 ± 1.8
B10.BR $\pm$ B10.BR$_m$	-15.4 ± 8.1	-18.9 ± 6.8	-16.8 ± 3.7

* % CML $\pm$ Standard deviation

Table 9
A "Three-Cell" Experiment

MLC			% CML		
Responding (Effector) Cell Donor	Stimulating (Sensitizing) Cell Donor	MHC Difference	Target Cell		
			AQR	6R	B10.A
AQR	6R	LD	-31.2	-6.6	-11.9
AQR	B10.A	SD	-31.9	-14.1	-7.1
AQR	6R + B10.A	LD + SD	-20.1	4.7	102.0

some of these reactions, as has indeed been suggested by certain investigators.

As already mentioned in the introduction, the Ir genes control the ability of an animal to respond to certain specific antigens. I will
not discuss these loci further at this time.

Molecular Considerations: The methods which have been
and will be described at this conference for the isolation of membrane components need no further emphasis here. I would rather classify these
methods into two categories to give us perspective in relationship to the
tests I have discussed earlier.

One approach to the isolation of membrane components utilizes antisera against those components. The isolation of the components
with which the antisera react after solubilization of the membrane with
NP40, or other agents, limits us strictly by the method used, i.e., to the
isolation of those components which elicit antibody formation. This may
well represent the great majority of components, however we cannot be
certain of this.

The other approach is to use a more general solubilization
of membrane components. This can either be achieved by the use of reagents such as 3M potassium chloride, enzymes or other agents which take
cell surface components off the cell membrane, or by the isolation of the
plasma membrane with subsequent general solubilization with agents such
as chloral hydrate or others.

The goal with any of these methods should be to have a reten-
tion of biological activity so that testing in various biological systems is
possible.

It may be worth adding one other way in which membrane
components may be obtained, namely that of shedding. This is something
which has been spoken about a great deal but has not been used extensively
for the preparation of membrane components.

The testing of the various MHC components discussed earlier
can be achieved in a number of different ways. First those components
which lead to antibody production, and where the antibody is cytotoxic to
the cells carrying the component in question can be used in inhibition of
cytotoxicity protocols. If a cell is reacted with a specific antiserum which
would ordinarily lead to lysis of that cell, in the presence of complement,
then the solubilized cell surface product can be used to inhibit that cytotoxic
reaction. This has been extensively used and represents a very valuable
tool for the assay of membrane components. It does little, however, in
ascertaining the biological activity of those components as far as we understand them. The approach can be used for the MHC SD and Ia antigens; the
latter perhaps being identical with the LD antigens.

With respect to the in vitro systems discussed in the earlier
section, the membrane components which are isolated can be used either to

stimulate a response or specifically block it. For instance, the MHC LD antigens could be used in an attempt to stimulate a proliferative response in a lymphocyte culture test system. If no proliferative response is obtained, then the same antigen preparation could be used to specifically inhibit the MLC reaction in question (where the stimulating cells are taken from the same donor from which the "solubilized LD antigen" is obtained).

In attempting to obtain stimulation with the LD or SD antigens, if they do not stimulate in the solubilized form, they may do so after attachment to an insoluble carrier such as sepharose or autologous cells including erythrocytes or lymphocytes. Alternatively, such preparations could even be attached to membrane preparations such as vesicles.

The preparations can, of course, also be used to sensitize animals in vitro to assay for their biological activity.

Once components are isolated, it will be possible to obtain antisera against them in both animals of the same species as well as in other species.

It is more difficult to consider biological evaluation of a presumed T cell receptor determined by the Ir genes. Even if such a receptor exists, the test systems which could be used to evaluate its biological role are not readily at hand. One possible model which may be worthy of consideration is based on the elegant work of Julius Adler in bacteria. Recognizing the differences in the structure of the bacterial cell wall and the mammalian membrane, the extrapolation from Adler's work is still worthy of consideration. Adler has shown that chemoreceptors can be switched between bacteria following their isolation from the membrane and that bacteria will manifest a phenotype consistent with the receptor rather than the genome which they carry following the switching of such receptors. Once again, whereas the technology may be different in bacteria and eucaryotic cells, this model may be a most attractive one for the biological evaluation of receptor molecules.

General Considerations: It is very likely that the many genes of the MHC to which I have referred in this brief discussion are the result of genetic duplications in evolution. I have spoken of the genes as having differential functions, such as the MHC LD and SD genes, and I believe that such differences are conceptually valuable for the time being. On the other hand, it is easy to imagine that in fact a large number of the phenotypic products of MHC genes are very similar and that dichotomies such as the one referred to between LD and SD (MLC and CML) will pale when we understand these phenomena at the molecular level.

One attractive model to explain the LD-SD dichotomy, for instance, would involve hypothesizing that the LD and SD gene products are both represented on the membrane but that the LD product has a much more rapid rate of shedding (turnover) than the SD product. As such, LD may function very well to stimulate T lymphocytes to proliferate but be a very poor "target" in CML (since the cytotoxic lymphocytes cannot attach to the target via the LD molecules). SD may be more firmly fixed in the membrane and, for some reason which is not clearly understood, may not be as good an activating substance for proliferation in MLC; on the other hand, the SD molecules may serve as an excellent target in CML.

   A last note which certainly must be brought to light is that the region we are considering, while a very small portion of the entire genomic material, is large enough, depending on one's assumptions concerning the percentage of genetic material in the human genome which consists of structural genes, for between 100 and 1000 genes. It may thus be that there are many more genes which need identification -- in turn perhaps some of these will determine other membrane components.

KAHAN: Thank you for that overview of the problem. If these genetic sites function via direct (or indirect) determination of surface components, one would predict several molecules controlled by the MHC system-H-2K, H-2D, Ss, LD, and Ir products. As Fritz pointed out, the analysis of these products depends upon the availability of specific biologic reagents-congenially bred mice, alloantisera, homozygous stimulator and responder cells, specific challenge antigens. Chemical and biological analysis of soluble membrane components with SD activity (H-20 and H-2K in the mouse, LD and FOUR in man) by immunogenicity for allograft rejection, inhibition of cytotoxic sera and induction of immunologic unresponsiveness has been extensively reported and will be discussed later. Chemical and biological analysis of LD (Ia) surface components is only at an early stage, due to the very recent identification of appropriate congenic lines and identification of sera recognizing this allotypic system. As Pincus pointed out in the previous session there is no evidence that antigens solubilized with sonication, KCl or papain and known to possess SD activity, do also possess LD activity. Indeed, Pincus' data on the capacity of surface-reactive reagents to abolish the MLC stimulator characteristics of a cell suggest to me that a more extended conformation on the surface may be required for stimulation of proliferation. This conformation may be more difficult to separate from the bulk membrane. The use of sera which recognize determinants within the allelic system Ia (which many people assume is equivalent to LD) may provide insight into allotypy of one portion of the Ia gene product. Dr. Sundharadas has devised another valid approach--examination of the protein component profiles of surface membrane derived from congenic mice which only differ by the Ia loci.

SUNDHARADAS*: As Bach already pointed out, a biochemical approach may indeed help us identify, purify, and characterize the membrane components controlled by the genes of the major histocompatibility complex. In order to take a rigorous biochemical approach, what is needed is a powerful method to disaggregate the membranes and analyze the components.

Sodium dodecyl sulfate (SDS) containing buffer systems have been used to dissolve membranes and analyze the components by gel electrophoresis. These solvent systems allow the separation of protein molecules on the basis of size only and different molecules of the same size cannot be separated. Furthermore, as we heard from previous talks, SDS allows aggregation of membrane proteins. Because of these difficulties with SDS containing systems, we looked for a method that would allow easy solubilization and disaggregation of the components and also would allow separation on the basis of charge.

Chaotropic agents offered a good possibility. However, chaotropic agents such as urea and guanidine hydrochloride are not powerful enough to disaggregate membranes. We examined several other compounds and succeeded in finding one, namely chloral hydrate, which we think is a very powerful chaotropic agent. We find that choral hydrate, 100% (W/V), also containing approximately 0.2M taurine and approximately 0.04M 3-bromopyridinium lactate (pH 3.0), is a powerful solvent for several biological membranes tested. This solvent system completely disaggregates the components of membranes, such as red blood cell membranes, plasma membranes of mouse thymocytes, spinach chloroplasts, and beef heart mitochondria. When these disaggregated membrane components are subjected to gel electrophoresis in the same buffer system, all the proteins enter the gel, get separated and give distinct and sharp bands. There is no sign of aggregate formation, which is often the problem with other solvent systems used to disaggregate membrane components.

The next question is, can we use this sytem to separate molecules on the basis of charge and not solely on the basis of size? The answer is "yes". The choral hydrate-gel electrophoresis method is capable of separating aminoethylated and carboxamidomethylated (-SH group derivatezed) myeloma light chain, MOPC 321. These two derivatives which differ in charge but have the same size, on the other hand, are not separated by SDS-gel electrophoresis. Similarly, two different light chains of the same chain length but different amino acid composition are separated by chloral hydrate gel electrophoresis but not by SDS-gel electrophoresis.

---

* The findings I am reporting here were possible primarily through the efforts of Dr. Bryon Ballou.

We have used this new method to compare the membrane components of thymocytes derived from two closely related strains of mice which apparently differ for only a short region of the chromosome within the major histocompatibility complex. The two strains are (B10A (2R) and B10A (4R). These strains are identical for the serologically detected H-2K and H-2D antigens, but one stimulates the other in mixed leukocyte culture reaction. A difference in 2 protein bands, 1 major band and 1 minor band, has been observed. These differences very likely represent allelic differences between the two strains.

A similar difference in 2 protein bands has been observed when the membrane components of thymocytes of another closely related mouse strains are compared. These are strains C57 BL/6 and H(21) which are thought to differ for a short chromosomal region within the major histocompatibility complex. They are identical with respect to H-2K and H-2D antigens but stimulate each other in mixed leukocyte culture reaction. These results indicate that chloral hydrate-polyacrylamide gel electrophoresis is capable of differentiating between allelic products expressed on membranes.

Another question that can be asked about this system is whether chloral hydrate damages or alters the proteins. Several pure proteins and amino acid mixtures exposed to chloral hydrate could be recovered unmodified. Also, the ribosomoal proteins of E-coli after exposure to chloral hydrate for 24 hours and $0^{0}C$ could be recovered with full biological activity. Finally, different lines of evidence suggest that purification of membrane proteins after disaggregation in chloral hydrate is possible and it looks like a very promising solvent system. In closing, I must mention that it is my judgment that technology is now becoming available to allow us to proceed with the study of membrane proteins.

UHR: We have been working on the partial characterization of the Ia antigens in the mouse, that is, the antigens referred to by Fritz Bach as the LD. These are studies done in collaboration with Dr. Klein. Our conclusions are in close agreement with those reached independently by Schreffler and Nathenson, and by Delovich and McDevitt, studying sera produced by immunizations against mice congenic for the Ia locus. The molecules which function as the antigens inhibiting these sera, have a molecular weight quite different from H-2 alloantigens, namely, about 35,000, as judged by mobility in SDS acrylamide gels. In addition, one can precipitate the H-2D and H-2K antigen from the labeled cell lysate, and not affect the specific immuno-precipitation of the Ia antigens of the gene complex Fritz Bach discussed. It is clear that the tissues which are most yielding of Ia antigens are those that are rich in B cells. It has not been excluded that the antigens are also synthesized by or present on the surface.

I think there is general agreement that we probably are not studying the products of the Ir genes, but there is the general suspicion that these may be the antigens concerned with stimulation in the MLC, and perhaps with other functions of a recognition nature.

There are two other additional findings of interest which we have made: One, a significant amount of Ia antigen was present in the incubation medium of tritiated leucine labeled splenocytes. The question, therefore, is raised whether this antigen might be actively secreted, shed from cells, or, of course, might come from lysed cells in the culture medium.

(I might mention as an aside that during this meeting, the term "shedding" is used quite casually by everyone to describe antigens known to be present on cell surface and to be found in some incubation media or, in some circumstances, in the serum. I think it is very diffi- cult task indeed to determine the origin of such material. Is it really secreted by a metabolically active transport process, or is it in fact shed by viable cells if that occurs in vivo, as it certainly does in vitro, or is it simply a product of lysed cells?)

The second finding is that normal sera frequently precipi- tated a small amount of radioactivity that migrated in the same area of the gel. This raises the possibility that the Ia antigen might have an affi- nity for immunoglobulin and in particular that it may be the Fc receptor.

KAHAN: Before you leave, I would like to ask how the membranes were prepared and were they then directly applied to SDS gels?

UHR: In these studies, cells were either labeled with tritiated amino acids or the cell surface was iodinated enzymatically. The cells were lysed with non-ionic detergents and immuno-precipitation was carried out in the presence of detergents; then the immuno-precipitate was solubilized in SDS.

WERNET: Can I just ask Jonathan Uhr whether he has found any evidence for beta-2 microglobulin associated with his Ia antigens?

UHR: We have not consistently found a peak in the area of beta-2 micro- globulin, but to throw in a tid-bit, we have found beta-2 microglobulin with another antigen also from this group nine linkage complex in the mouse, namely, the TL antigen.

WERNET: Several ways have been explored so far in order to define the membrane products of the major histocompatibility complex, and particu- larly those which are involved in the generation of blast transformation in vitro, particularly in the mixed leucocyte culture. The work described here has been done in collaboration with M. Giphart, A. von Leenwen, P. Wernet, H. G. Kunkel and J. J. von Rood.

The system we have been trying to use there was the inhibition of the MLC by alloantisera. This approach had been used by Cepellini and his collaborators, and later by Eric Thorsby and his group, who were able to quite successfully attribute the blocking effects to anti-HL-A sera. It is clear, however, that the HL-A antigens do not represent the recognitive structures, and also are not the primary target-structure for immunological recognition. So how can their results by explained?

Anti HL-A sera can discriminate different targets on different lymphocyte membranes besides the HL-A antigens, most likely, because they also have reactivities which are directed against other antigens than HL-A. Those antisera are candidates for containing antibodies which have specificity for the product of the mixed lymphocyte culture controlling genes, if inhibit such interactions, particularly in HL-A identical unrelated combinations.

In contrast to such cytotoxic anti HL-A sera, we have been looking into the possibility of utilizing such sera which were not cytotoxic with complement, but which showed a considerable positivity on certain lymphocytes with indirect fluorescence (15 - 25% on peripheral lymphocytes), and also at the same time, were able to specifically inhibit the stimulatory function of these lymphocytes in MLC-combinations. Several of such sera proved to follow specific genetic lymphocyte-types, as assessed by HL-A and MLC-typing. (A. von Leenwen or J. J. von Rood, Transpl. Proceed. 5, 1973). The MLC-typing correlated well with positive indirect immuno-fluorescence with anti IgG.

On the first figure (Figure 50), you see an example of such antiserum which derived from a multiparous woman (PC) discriminating its targets on the membrane of lymphocytes Ma, which are blocked in their stimulatory MLC-function by this sera. The HL-A type of these lymphocytes is identical to that of the serum donor (HLA-2, 12, 9, W5). Depicted are sliced SDS-arylamide electrophoresis gels of radiolabeled and NP-40 solubilized Ma-lymphocyte-membranes.

You see in A, the main area is about at position 22, and another peak at 32; and if the molecular size of these components is estimated, it will possibly be close to 80,000, and the other peak close to 30 to 40,000 daltons, in addition a small peak of 11,000. Since this pattern is fairly close to that obtained for the HL-A antigens, it had to be ruled out that this still could be an antibody directed against histocompatibility antigens which is just not cytotoxic.

The midportion (B) of the figure shows that iodination profile of Ma lymphocytes obtained with the antiserum P1, which has been absorbed with Ka-lymphocytes, whose stimulatory capacity also was inhibited by the original serum T1. It can be seen, that in B, the peak of

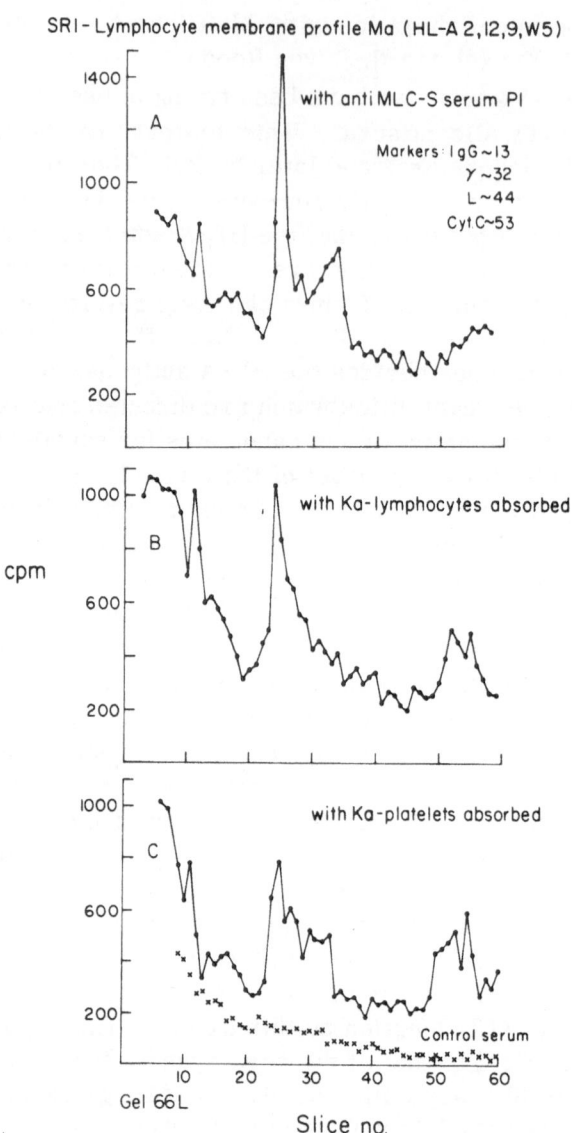

Fig. 50. Surface radioiodination profile of Ma-lymphocyte-membranes, (A) precipitated with serum Pl, which inhibits the stimulatory function of Ma lymphocytes. In (B), the serum Pl was preabsorbed with lymphocytes Ka, whose stimulator-function also was abolished by the Pl serum prior to absorption. One major peak is not discriminated any more and seems to be a candidate for an MLC-S locus-product. In (C) serum Pl had been preabsorbed with Ka-platelets, in order to test for hidden anti HL-A activity. All major peaks are still identified and this anti-serum still inhibits Ma's MLC-S.

30-40,000 daltons is not present any more, which would mean that the antibody specific for this target has been absorbed out. The peaks at about 80,000 daltons and 11,000 daltons are still present in this picture.

Quite in contrast to this is the bottom part of this figure. This shows the surface iodination profile after the serum T1 had been absorbed with aged platelets of Ka, which carry the matching HL-A antigens. It can be seen that all peaks are still present. There seems to be a slight reduction of the peak at position 22 and in the small molecular weight area, a slight increase can be noted.

This complex pattern may provide the information, that the MLC-S structure is a complex one which would contain also (like HL-A) beta-2 microglobulin and perhaps also appear in dimeric form on the lymphocyte membrane or in association together with HL-A antigen. A model of a tandem associated between HL-A and MLC-S can be discussed where beta-2 microglobulin would provide the basic carrier molecule on the membrane. One HL-A specific polypeptide chain could be hooked up to one beta-2 microglobulin and one MLC-specific polypeptide chain to another beta-2 microglobulin. Both the HL-A and the MLC units could be held together in close unity by carbohydrate parts. In this way, one could also understand how specific anti-HL-A antibodies would be effective in inhibiting MLC by steric hindrance.

Another question concerns the primary recognition in the mixed leucocyte culture, which seems to be a function of the T cell whereas stimulation in MLC is probably not mediated by T cells.

Here it was possible to utilize serum from patients with systemic lupus erysthematosus (SLE) which were able to block allogenic lymphocyte stimulation and response in a non-specific fashion as well as specific through antibody. The isolated IgG antibodies from certain patients was particularly effective in blocking the MLC response of the autologous serum donor, when tested with unrelated HL-A different and HL-A identical combinations without being cytotoxic. In order to define specific membrane structures via homologous antibodies associated with MLC suppression, lactoperoxidose catalyzed surface radioiodination (SRI) of lymphocytes has been employed. In several experiments, the peripheral blood lymphocytes were isolated further after Ficoll Hypaque separation. Passage through Sephadex G 10 columns eliminated macrophages almost completely. Further passage through nylon wool columns resulted in highly purified T cell populations. Using as precipitating antisera in SRI carefully absorbed MLC-R blocking SLE derived IgG antibodies, the following results were obtained. Figure 51 shows in the upper half the SDS-polyacrylamide gel electrophoresis pattern, which was obtained with radiolabeled SLE lymphocyte-membranes, precipitated down with autologous serum antibody. Several target structures are identified. From these patterns, the impression was gained that active SLE antibodies of the IgG

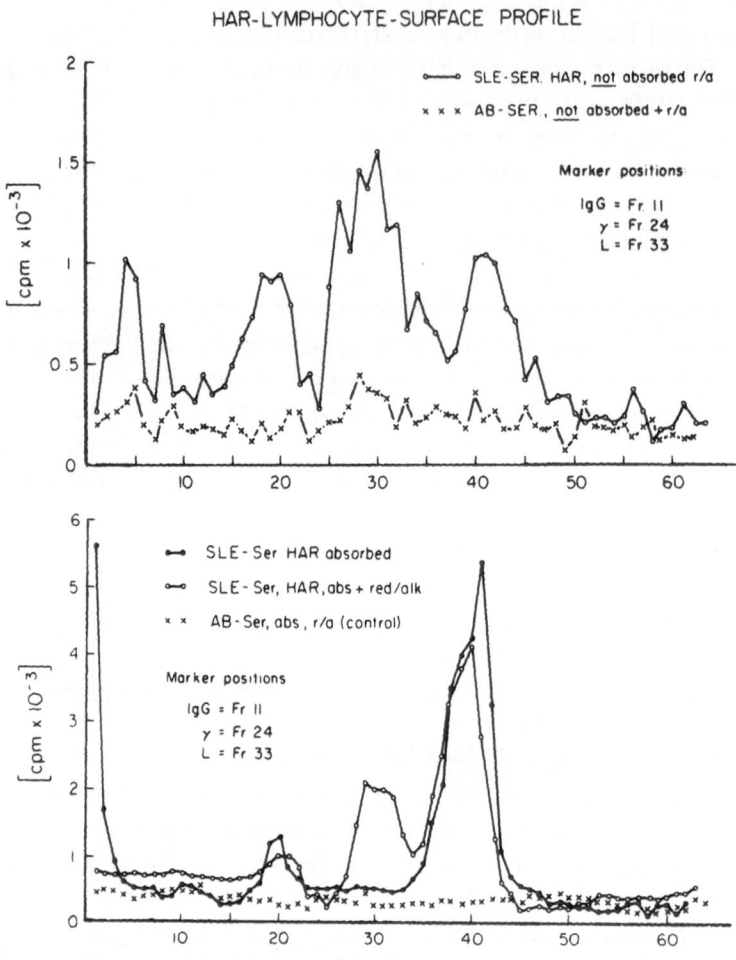

Fig. 51.   SDS polyacrylamide gel patterns, representing target surface radioiodination profiles obtained from membranes of SLE peripheral blood lymphocytes using the corresponding autologous SLE serum antibody (IgG) as the antiserum to precipitate the membrane components.   Lower part shows patterns obtained with the absorbed antibody which still inhibits the MLC response.

type contain a whole group of antimembrane antibodies, directed against different lymphocyte-surface structures.   The lower part of the figure shows in the curve with the black circles the pattern obtained with absorbed IgG from SLE-serum Har.   The absorption was carried out with ABO red cells, pooled platelets and "B"type lymphocytes from a patient with chronic lymphocytic leukemia.   This absorbed antibody was still able to block the MLC-response of the autologous serum donor cells and thus can be called an autoantibody against immune recognition. Besides a small peak at about 80,000 dalton, one major peak of about 15,000 to 17,000 daltons is identified.   If the same precipitate, however, was reduced and alkylated before electrophoresis, the gel pattern changed. This is shown with the curve with open circles.   The main difference is the appearance of a strong peak at position 30, which would be in the area of the major polypeptide chain of HL-A antigens (see Wernet, these Proceedings). Thus the molecular arrangement for the recognition part on the lymphocyte membrane, which seems to be involved in the MLC-response, looks complex. Also, there is a specific piggyback arrangement on the basis of beta-2 microglobulin possible.   Also, a molecular side-to-side arrangement with HL-A seems possible.

Because of the protein denaturing properties of the SDS-gels, another method to isolate such receptor structures would be welcome. In this regard, an attempt is made to utilize the shedding of such structures into culture medium. This phenomenon is particularly strong after the lymphocyte membrane is coated with the matching antibody. Figure 52 shows such an experiment. Isolated T cells were radiolabeled coated with anti "MLC" antibodies and cultured for 12 hours in the presence of Trasylol[R], which serves as proteinase-inhibitor. The supernatants were concentrated and run on Sephadex G200 columns at pH 7.6 and pH 3.0 respectively. The black points symbolize measured radioactivity. The comparison of both runs demonstrate the existence of shedded membrane components, complexed to the antibody. At acid pH, the complexes have fallen apart and disclose as candidate for the shedded membrane receptors mainly two peaks in a smaller molecular weight range. This way to study membrane structures has to be explored further since it may be a most useful tool to characterize such receptors or at least their single molecular components.

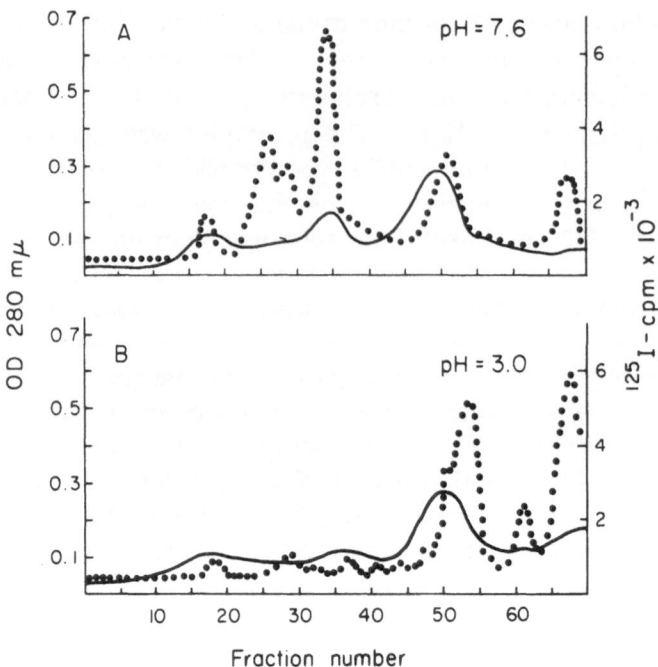

Fig. 52. Shedding-profile of radiolabeled, anti-MLC-R coated and cultured lymphocyte supernatant after fractionation on Sephadex G 200. Black points represent radioactive counts.

KAHAN: I had just one question for John Uhr, and I am sure it is something that crossed everyone's mind since Fritz introduced this topic Can one be totally happy to use antibodies to study antigens whose determinants are presumably discriminated by a cellular immune process. That is, if proliferation is a phenomenon related to cellular recognition -- in other words, one cell contacting another cell and dividing -- what assurance is there that the allotypy recognized by humoral antibody is discriminated in the same fashion as this cell-cell interaction.

UHR: I am not concerned about the problem, because in this case, antibody is simply a very convenient probe to isolate the moleculte. In fact, this is the only probe we have at the present time. I think what you are getting at is when we have the molecule isolated, how is one going to relate that to the phenomenon of recognition at the cellular level, particularly if the recognition unit is other than antibody? I think that is something that will have to be worked out at another stage.

KAHAN: Do you want to say something about the relationship between Ia and MLC, Fritz, so we don't get caught in jargon?

BACH: The MLC antigens, and the LD antigens are synonymous terms, just different terminologies that are being used.   But the important question which Barry raised to John Uhr is essentially, can one hope to define the MLC and LD components with antisera. I think there is a very good likelihood that they are immunogenic for B or bone marrow derived cells and one will get antibodies to them. It seems at the moment, by these test procedures, that the SD antigens are preferentially stimulating antibody production, and the LD antigens are preferentially stimulating a T proliferative response. It may be that the Ia antigens are the same as the LD, however, it is very important in dealing with a chromosomal segment like the MHC not to reach the conclusion that the Ia antigens are the products of the LD locus.

KAHAN: Can we take the same animals that have these gene differences and detect -- rather than humoral antibody, evidence of cell mediated immunity --

BACH: One knows that there are several different gene products, those which Uhr and Wernet spoke about, and those analyzed by Sundharadas, reflecting multiple differences between these strains with respect to the MHC, not only the H-2 sites K and D.   I think that it is a mute question whether all these products will cause antibody production in alloimmunizations or whether some like LD, will lead only to a T lymphocyte proliferative response.

McKHANN: It seems that each of the SD and LD components on the cell surface, may really exist in two forms.   One is a "receiving" form and the other is a "sending" form. One is the receptor which must receive information, and the other is an effector site which can give information that brings about cell stimulation or cell killing. With respect to cell killing, the effector site is presumably antibody, or an antigen specific site on a lymphocyte. But I don't think that we know that much about the "sending" site with respect to the LD complex. Do you think these may exist as two separate conformational areas on the cell surface, each with its own specialized function for receiving or sending information may then serve a double purpose, particularly the lymphocyte recognition site.

BACH: I think the first thing to remember is that immunoglobulin, which is the classical receptor in immunology has idiotypes, antigens, which are recognized by antibody molecules as foreign. I am presuming that cell surface structures determined by this region, such as the LD and SD antigens are receptors, and also will manifest as antigens. The question has been raised whether the same molecule functions to provide the antigenic stimulus in mixed leucocyte cultures as the one which recognizes the foreign antigens.   I don't think there is critical evidence in that regard. Certainly, the response of mixed culture is controlled to some extent in this region, but that is very hard to separate out. At least I think we have to hypothesize a separate antigen system.

McKHANN: Could you tell us what evidence there is for or against a con- tinuity of responsiveness in terms of the cells that are stimulated and those that become cytotoxic lymphocytes. Is there any evidence these are one and the same cell.

BACH: I didn't think this is related very much to the membrane, so I left it out.   The question is whether the same cell which responds to LD and proliferates, which we assay by studying the incorporation of thymidine, also differentiates to become an effector, or killer, cell.
          Let me say that I don't think there is a final answer on this, but there is some evidence, which is consistent with the concept that we are dealing with two thymus-derived lymphocyte populations, one of which proliferates and responds primarily to LD, and a second subpopu- lation of thymus derived lymphocytes which become killer cells. And, that the primary target of this second population, the cytotoxic target, is the SD antigen, or the phenotypic product of genes very closely linked to those determining the SD antigens.   I don't think there is critical evidence in that regard.   I think if that is not the case, that they are not two popula- tions, then we have to say probably that whatever cell recognizes LD and starts proliferating, undergoes an additional differentiative step to become a killer cell when it sees SD.   This would require one cell to have two, I think, separate recognition systems, both clonally distributed.   I don't think that is impossible.

WERNET: I just have one thing which is not clear about target cell gene- ration, in cell mediated lympholysis.   Since the target cells there are generated by PHA stimulation and PHA is added in a soluble form, it would generate only T cells and not B cells to be transformed into blasts. The primary stimulatory antigen, however, seems not to be present on the T cells, so I wonder whether you could perhaps get a positive killing res- ponse in SD identical, LD different populations if you would use other sti- mulation systems, for example lipo-polysaccharide in order to generate B lympho-blasts as target cells.

BACH: The question is a very good one.   The target cells labeled with sodium chromate and are stimulated with phytohemagglutium.   The argu- ment could be made that these stimulated cells don't carry LD and that is why LD does not function as a target.   Please let me stress here again that we think that LD may function as a very weak target, much weaker than SD.   Barbara Alter and I have looked at PHA stimulated target cells, Con A stimulated target cells, LPS stimulated target cells, and because some people claim that maybe the same cells carry the recognition sites for MLC and the LD antigen, mixed culture stimulated targets, and the most critical normal fresh lymphocytes obtained from the lymphnodes of animals as target cells, and in all those cases the results are identical.

KAHAN: We associate two steps presumably based on the activity of two different cell types. (At least that would be the explanation I would favor.) Some cells proliferate in response to stimulation, other cells probably differentiate and display immune performance. The proliferative type of cell can be detected in mixed leukocyte cultures, which is the interaction of non-sensitized cells.   Two normal cells are mixed and the response obtained is regulated by the MLC (LD) products which have been discussed thus far.   In contradistinction to this proliferation of unstimulated cells, cells which have been previously sensitized toward SD antigens can display specific proliferation when confronted with the corresponding SD antigen.   This antigen specific mitogenesis is similar to that displayed by tubercle-sensitive lymphocytes upon contact with tuberculin.   The other state, namely immune performance, may be much more important. Performance involves the generation of effector cells capable of killing targets, and the release of mediators of the immune response.   All of these phenomena might be potentially dissected by in vivo administration or in vitro exposure to soluble material. The frustration today is that the soluble materials which have been employed in serological analysis have not yet been shown to be nearly as effective in mixed lymphocyte culture reactions or in cell-mediated cytotoxicity phenomena.   However, the fact that some soluble products can inhibit the action of cytotoxic effector cells suggests that the molecular analysis of this form of cell mediated immunity is within sight.

FERRONE: One of the reasons why soluble materials have been more extensively analyzed by serologic techniques is that we have a better understanding of the mechanism of antibody-mediated complement-dependent cytotoxic reactions. For example, we are able to study the role of surface expression of SD antigens during different phases of the growth cycle. The experiments which I am going to present were done in collaboration with Drs. Pellegrino and Cooper and utilized the cultured human lymphoid cells RPMI 8866. The results obtained indicate the following: 1) the cells change in susceptibility to lysis during different phases of their growth cycle: 2) cells in G_1 are least sensitive to immune antibody-mediated lysis; 3) there is no correlation between changes in their susceptibility to lysis and the expression of HL-A antigens, the binding of antibodies or the activation of the complement system.

When we utilize cultured human lymphoid cell, RPMI 8866 as targets in the complement dependent cytotoxic test, we observed that the cells in the Gl phase were less susceptible to lysis mediated by HL-A antibodies directed both against antigens of the first and of the second segregant series. The results were consistent when we utilized different sources of complement such as rabbit, human and guinea.pig sera. The latter two sources were utilized to exclude that the change in susceptibility to lysis reflected changes in the amount of antigens against which the natural antibody present in rabbit serum was directed. The lysis of target cells in the complement dependent test is the result of the interaction between antigens, antibodies and complement, and therefore, we tried to solve the question whether this change in susceptibility to lysis was reflected by changes in the expression of antigens, in the binding of antibodies or in the activation of the complement. In order to quantitate the expression of HL-A antigens on the surface as the cells move throughout the cell cycle, we utilized essentially three criteria -- one was the radioactive antiglobulin test which consists in reacting the cells with the anti HL-A antibodies, and then quantitating the binding of the antibody by reacting with radioactive rabbit anti human gamma globulin serum. When we performed this test, we observed that there was no significant change in the amount of binding between cells in Gl and in the Go phase both for HL-A 2 and for HL-A 7 specificity. There was an increase in the amount of radioactivity, when we utilized cells in their S phase. In order to express these data in terms of density of antigens on the cell surface, we had to correct the values according to the volume of the cells as they move throughout the cell cycle. By determining the volume of the cells, we could observe that cells in S phase had larger volume than cells in Go and Gl, and this difference in volume could account for the difference in binding of radioactivity we observed. Similar results were obtained when we determined the binding of the antibodies by the cells by utilizing a quantitative microabsorption test as well as when we determined the amount of antigen present on the cells by measuring the yield of soluble HL-A antigens from the cultured cells by the 3M KCl method. We had to reach the conclusion that there was no change in the expression of HL-A antigens, and there was no change in the ability to bind HL-A antibodies as the cell moved through the cycle.

The next question was, does the amount of complement components bound by the cells change as the cell moves through the cycle. We determined the amount of labeled C4 and C3, which represent the early complement components, and the amount of C8, which was bound by cells in Go, Gl or S, when they were coated with HL-A antibody, or with anti-human lymphocyte serum, or with rabbit IgM. As it appears from the data, there is no change in the amount of complement components which were bound by the cells as the cell moved throughout the cycle.

The next question was: does the activation of the complement system change when the targets are in different stages of their growth cycle. I would like briefly to outline how the complement can be activated. I will try to be very simple. Those who are familiar with the complement, will forgive me for this simplistic description of the complement system. There are two pathways of activation of the complement. One, the classical pathway which starts with C1, proceeds with the formation of the complement, C4, C2, which cleaves C3 and then the activation proceeds until the activation of C9. The other pathway is a more recently described pathway which is called the alternate pathway, or is referred to as the properdin pathway system. It is not yet as well defined as the classical pathway. It starts with the properdin and then a series of factors which join to the classical pathway, at the level of C3, which cleaves C5 and then the reaction proceeds to C9. It is possible to differentiate between the activation of 2 complement systems by several means, and we utilized essentially two which I am going to describe. The first one takes advantage of the fact that the classical pathway requires for its activation the presence of calcium in the medium. Therefore, by chelating the calcium with EGTA, it is possible to inactivate the classical pathway, while the alternate pathway will still function in the regular way. The second approach took advantage of the fact that it is possible to destroy a factor of the alternate pathway, the C3PA by simply heating the complement source at 50 degrees for 20 minutes, and then it is possible to reconstitute this pathway by the addition of purified C3 PA. There is no change in the activation of the complement system by cells which are in a different stage of growth cycle. Specifically, utilizing antibody, which proceeds through the activation of the alternate pathway, there is no lysis when the alternate pathway is inactivated by the destruction of C3PA. Lysis is reconstituted upon addition of the purified C3PA. On the other hand, the lysis is not affected by the addition to the mixture of EGTA, which blocks the classical pathway without affecting the alternate pathway.

Therefore, the results indicate that it is possible to obtain cultured cell lines which change in the susceptibility to lysis as they proceed through the growth cycle. This change in susceptibility to lysis does not reflect any change in the amount of antigen, in the ability to bind antibody, and in the ability to bind as well as to activate the complement system. Therefore, the suggestion follows that this change in susceptibility to lysis reflects changes which occur in the cell membrane as it proceeds through the growth cycle.

KAHAN: How is the inhibition analysis performed specifically? How are soluble antigens assessed for HL-A activity? Can the soluble material be used to produce antibodies which may be tools for dissecting HL-A antigen?

FERRONE: The soluble HL-A antigen is usually tested by its ability to
bind HL-A antibodies in the complement dependent cytotoxic test, and
therefore, to inhibit the lysis of selected target cell.   Barry asked me to
outline some of the problems which everybody has to face when utilizing
the inhibition test to evaluate soluble HL-A antigens. The problems mainly
derive from the fact that the source of HL antibodies is unfortunately very
poor.   The antibodies are usually obtained from pregnant women or from
subjects who are immunized by blood transfusion.   When we test antigens
for their ability to block the cytotoxicity of sera against target cells.   The
various targets require different amounts of antibody.   This variable is
very easy to control; unfortunately, there are other variables which are
more difficult to control, and I have outlined these variables.   One variable
focuses on the effect that different HL-A alloantisera directed against the
same HL-A specificity vary in their susceptibility to inhibition by soluble
HL-A antigens.   (Figure 53)   and Table 10.

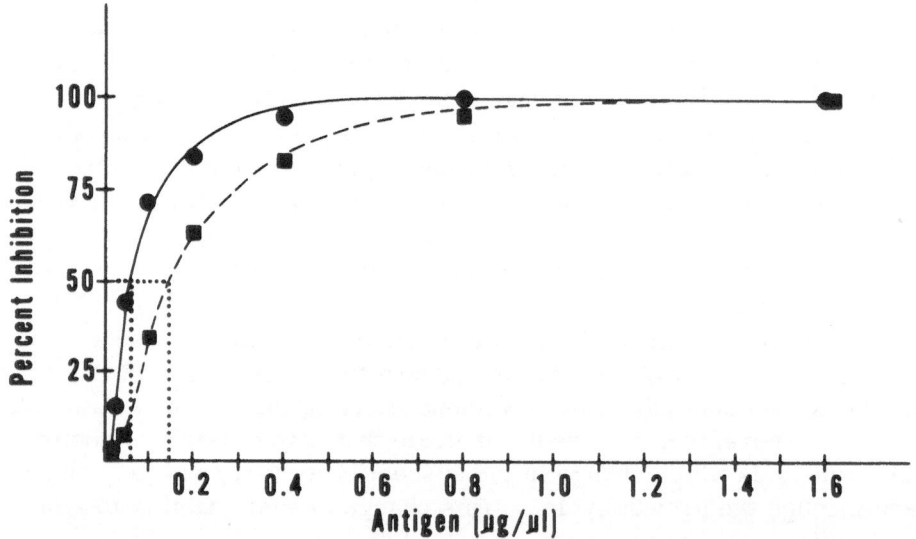

Fig. 53. Influence of the titer of HL-A alloantiserum utilized in the inhibition test on the activity of soluble HL-A antigen extracted from cells RPMI
788 (■—■) serum utilized at 2 cytotoxic units, (●—●) serum utilized at 1
cytotoxic unit.

The next figure shows that the target cells can be a source of variability.

Table 10

Different Susceptibility of HL-A Alloantisera to Inhibition by Soluble HL-A Antigens

anti-HL-A2 antisera		Antigen #115-103 (2,10,7,W14)	anti-HL-A3 antisera		Antigen #96-103 (3,Te 63)	anti-HL-A5 antisera		Antigen #147-103 (1,2,5,Te 57)	anti-HL-A7 antisera		Antigen #129-103 (2,10,7,W14)
Name	Titer*	ID_{50}	Name	Titer	ID_{50}	Name	Titer	ID_{50}	Name	Titer	ID_{50}
To-11-03	32	0.03	Storm	32	0.035	D-66	8	0.13	Cutten	8	0.06
Eriksson	64	0.03	Tucker	128	1.00	McMullen	8	0.13	Melnikoff	8	0.06
Stokenberg	128	0.10	Kraska	4	1.00	Victor	128	0.13	Jackson	32	0.06
Pinquette	128	0.10							Cowen	16	0.13
Stewart	64	0.10							Haas	4	1.14

* expressed as the reciprocal

Table 11

Relationship Between the Titer of Cytotoxic HL-A Alloantisera
and Sensitivity of the Inhibition Test

anti-HL-A alloantisera			Target Cell #	ANTIGEN ID_{50}		
Name	HL-A specificity	Titer*		#115-104 (2,10,7,W14)	#96-104 (3,Te 63)	#147-104 (1,2,5,Te 57)
Stokenberg	HL-A2	32	3	3.20	-**	-
"	"	64	15	0.24	-	-
"	"	128	1	0.08	-	-
Storm - 22-02	HL-A3	16	15	-	.10	-
"	"	32	1	-	.03	-
D-66	HL-A5	4	15	-	-	0.75
"	"	8	1	-	-	0.15
Cutten-11-01	HL-A7	8	15	.26	-	-
"	"	16	1	.13	-	-

* expressed as the reciprocal   ** not done

For instance, if we utilize the same HL-A alloanti serum against the 5 different target cells, we'll find a different value of the activity of this soluble HL-A antigen. Unfortunately, the HL-A test and therefore, the inhibition test is not standardized and this will cause problems when somebody wants to compare results in terms of activity of antigens which are obtained in different laboratories, or sometimes when somebody in the same laboratory is obliged to change the target cells to be utilized in the inhibition test.

The last point is the immunogenicity of soluble antigens to induce xenogeneic hosts to produce monospecific anti HL-A sera. For those who are not familiar with the HL-A field, I should mention that there has been a dogma for quite a few years that it is not possible to produce specific cytotoxic HL-A xenoantisera. This dogma stems from experiments which were done by some groups between 1956-1967, in which attempts to produce specific HL-A antisera by immunizing rabbits with whole human lymphoid cells failed. We have tried not to be prejudiced in this regard, and we had to fight against the HL-A brigade who did not accept the idea that it is possible to produce HL-A antisera. In the beginning, we utilized HL-A antigens which were solubilized from cultured human lymphoid cells by the 3M KCL method. The results were quite disappointing and seem to validate the dogma that it was not possible to produce monospecific HL-A antisera.

More recently, we have been much more lucky, and the reason for this luck is that we utilize a different source of the antigen for the immunization. Specifically, we used soluble HL-A antigens, partially purified, not from cultured human lymphoid cells, but rather from serum. When we looked at the specificity of these antibodies, the first bleeding from the rabbit showed a multi-specific pattern. Later bleedings showed cytotoxic mono-specific antibodies at the appropriate dilution.

As I mentioned, we had to convince the HL-A people that it was indeed possible to produce specific antisera, and the only way to convince them was to send the antiserum to their lab and to have them test the antibody. We selected some of the labs which are involved in the study of HL-A antibodies, and we asked them to test the activity of this antibody in their test system against the target cells. The labs which tested the sera confirmed the specificity of the antiserum which we have sent them. The only exception was represented by Kissmeyer-Nielsen, who did not find any reactivity of this antiserum, but the reason for this is very easy to discern, namely, he uses a much less sensitive technique for the detection of HL-A antibodies.

KAHAN: Thank you very much. I think what we have put forth is 2 systems for working with histocompatibility antigens on the cell surface; one dependent on cellular activity, the other depending on humoral antibody. Not only is the action of specific tissue typing sera inhibited by solubilized derivatives of membranes but also these materials can induce the production of specific antibody useful for tissue typing.

McKHANN: This is a very interesting finding since this is one biological difference between antigen recovered from the circulation and that recovered from the cell membrane. Could you tell us how similar the circulating antigen and the surface antigens were when you have been able to recover both and compare them.

FERRONE: One thing which should be pointed out about soluble HL antigens in serum is that it seems certain that there is a difference in the amount of HL-A antigenic specificity which we find in serum and that which we detect on cells. Without going into too much detail, HL-A 9 present in large amount in serum, completely disproportionate to that represented on cells. I don't think we can make any statement concerning these differences in terms of physical chemical properties of HL-A antigens from serum and from cells, except perhaps for the fact that in Dr. Reisfeld's lab, it was shown that the HL-A antigens recovered from serum have the same molecular weight of those recovered from cells. In addition, the serum HL-A antigens do not seem to contain carbohydrate, a finding similar to that obtained with antigens prepared by 3M K Cl extraction of cultured cells. The only difference thus far is that HL-A 9 seems to have different physical chemical properties than antigens with other HL-A specificities. Unfortunately, it is not yet possible to make a comparison with HL-A 9 isolated from cells, since as far as I know, there is presently no cell line available which contains HL-A 9. We have, however, recently obtained one cell line which contains HL-A 9, and we are in the process of characterizing HL-A 9 solubilized from cultured human lymphoid cells.

KAHAN: I think in terms of the overall evolution of this discussion; we really can't say that the soluble derivatives are perfectly valid representatives of transplantation antigens as they exist on the cell surface, unless we can show that these fractions are capable of tolerance induction, since the antigen which is presented to the host must be a complete copy of that on the surface in order to obliterate responsiveness.

APPELLA: Together with Dr. Lloyd Law and in collaboration with Dr. Sam Strober and Dr. Peter Wright, I have studied the biological characterization of H-2 antigens. What we were really interested in was to try to combine a biochemical and a biological characterization of solubilized H-2 antigens. If I may have the board, please, I will outline the H-2 complex. In this complex there are two main regions, H-2K and H-2D, which apparently code for antigens detectable by both serological and transplantation methods. The products of these two subloci have been extensively studied by Nathenson and Davies and when we approached our study, we decided that among a number of ways of solubilizing the antigens, we should follow their method using limited papain digestion. We chose a strain, the A/J strain, as a source of our material. However, this choice turned out to be very interesting and important. H-2^a is the haplotype which consists of a number of serological specificities both private and public; private being specificities of the D or K end, and public otherwise.

Papain solubilized material from this strain may be obtained in a high yield and partial purification can be achieved by gel filtration on Sephadex G-150. The use of a lectin sepharose column and immunoprecipitation with alloantisera can increase the purification. The molecule appears to be a glycoprotein of a molecular weight of about 54,000, can be easily disassociated by acid in two subunits with a large component of about 33-36,000, and a small component around 12,000 that resembles the beta-2 microglobulin very much.

For the biological studies, we used the so-called F2 fraction and that is a partially purified material from a Sephadex G150 that has the molecular weight of about 54,000, I would say, and consists, as you know by now, of the 2 subunits. In addition, it has all the antigenic specificities that can be tested by using the available alloantisera. Table 12 is a summary of the most relevant and easily to understand biological characteristics of this solubilized and partially purified H-2^a antigen.

Table 12. Biological Characterization of F2 Fraction (H-2^a Antigen).

Yield 65% of CM; specific activity 32X that of CM

Specific activities confined to this fraction and peaked in same tube

Activity of allogeneic specificities = H-2.1, 3, 4, 5, 11, 23, 25, 28 (immunologically specific)

Transplantation immunity (skin graft and neonatal heart rejection) elicited in congenic mice by F2 and not F1 and F3 fractions

Cytotoxic and hemagglutinating ab induced by F2 in congenic B10.D2 and B10.B mice

Specific immunologic enhancement induced by F2 fraction (A strain tumors in congenic mice)

Humoral tolerance induced in congenic B10.D2 mice (specific and long lasting)

There is a yield of about 65% with about 32-fold purification for the fraction. All the serological specificities were accounted for and the transplantation immunity (skin graft and neonatal heart rejection) was elicited in congenic mice by this fraction. We did not find any activity in an Fl fraction that was at the exclusion volume or a very low molecular weight fraction, F3. If, on the other hand, noncongenic mice were used, you could find transplantation immunity in all fractions, so H-2 and non H-2 were solubilized. Cytotoxic and hemagglutinating antibodies were induced by F2 in congenic mice and a specific immunologic enhancement was also induced. This enhancement was specific and induced by active immunization, and passive transfer and demonstrated that all important antigenic determinants have been recovered in the soluble material. The last is the most critical study and it is tolerance. As it is written there, humoral tolerance in neonatals specifically was induced by the F2 fraction, but not complete tolerance. Complete tolerance (allograft tolerance) could be achieved by the use of intact cells, but not by the use of soluble antigen in the combination of strains B10.A - B10.D2 used in our studies.

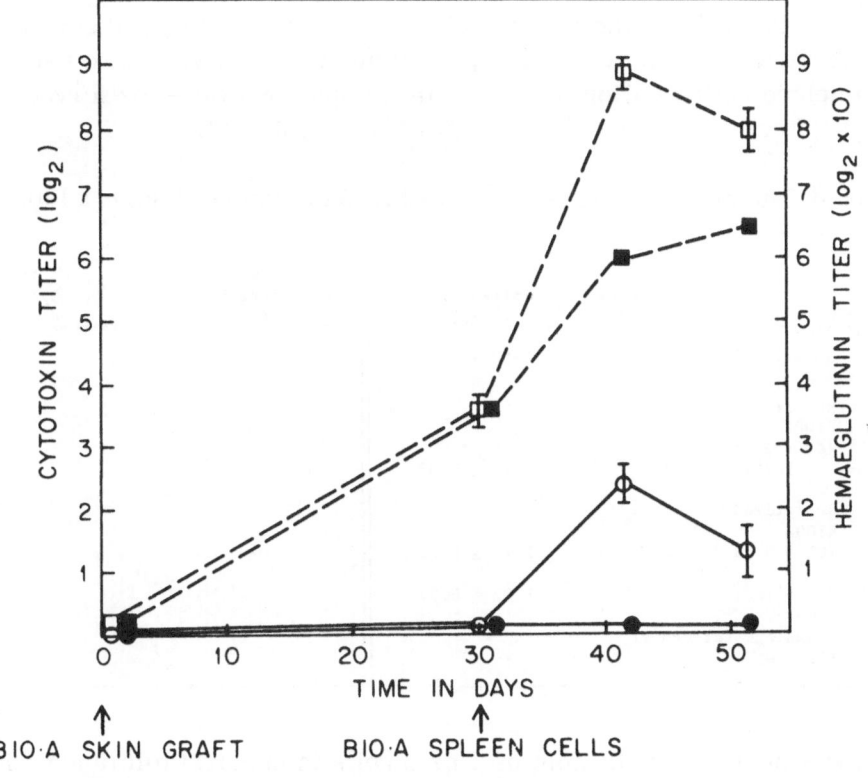

Fig. 54. Cytotoxic and Hemagglutinating Antibody Response of B10.D2 Mice after Skin Grafting and Inoculation of Spleen Cells (B10.2 donors).

        The protocol used for assaying cellular and humoral toler-
ance which resembles very much the one used by Mitchison and Nossal for
other protein antigens. Solubilized material was injected at day zero into
the heart. Then there were daily injections intraperitoneally for about 28
days, at which time skin grafting was done. Evaluation of results of the
time of complete rejection between control and the treated animals showed
no difference. Humoral response, however, was markedly depressed and
nonexistent in the majority of treated mice. Figure 54 shows that antibody
titers were undetectable.

        Antibody responses were measured again at 11 and 21 days
after challenging with B10. A spleen cells; still there was a big difference
in the control and the treated animal.

        If both groups of mice received B10 spleen cells and then
antibody responses were measured (we are looking now at the $H-2^b$), there
was no difference between the control and the treated animals showing
specificity of response. This humoral tolerance lasted quite a number
of weeks. For some animals as long as 16 weeks without further administration of the soluble F2 antigen. We have also assayed the reactivity of
lymphoid cells from humorally tolerant mice; specifically the graft versus
host reaction (GVH), the mixed leukocyte reaction (MLR) and cell mediated lysis (CML). Whereas GVH, MLR and CML activity were not found
in lymphoid cells of allograft mice, these functions were preserved in
lymphoid cells of humorally tolerant mice. (Table 13)

Table 13. Summary of GvH, MLC and CML Reactions in (B10. A x B10. D2)F_1
Mice.

Group	No. Spleens Assayed	Spleen Index (± SE)	Mixed Leukocyte Reaction (MLR)	Lymphoid Cytotoxicity (CML)
1. Normal B10D2	14	1.52 ± 0.14	+	-
2. Sensitized B10D2 (CT = 32 to 1024)	14	1.70 ± 0.03	+	+
3. Sensitized B10D2 (CT = 0 to 2)	16	2.11 ± 0.13	++	+
4. Sensitized B10D2 (Complete tolerance) (CT = 0)	9	1.02 ± 0.07	-	?

Data was obtained on the role of T or B cells in humoral tolerance. Neo-
natally thymectomized animals were reconstituted either with thymocytes
or lymphnode cells from normal or humorally tolerant animals; then

challenged with antigen in the form of intact (B10. A x B10. D2) F1 c e l l s. There is a significant difference in the cytotoxic titers with the lymphoid cells derived from humorally tolerant animals versus the normal, so that a subpopulation of T cells seems to be affected. No significant differences were found with the thymocytes. This part of our investigation is not yet complete so that we cannot as yet state with certainty where the lesion (T cell, B cell or both) is located. (Table 14)

An experiment carried on by Dr. Strober with reconstituted bone marrow irradiated animals showed that upon challenge with F1 cells there were no differences in the cytotoxic titers of treated and normal animals. This would indicate that B stem cells are not involved in humoral tolerance. More recent results obtained in collaboration w i t h D r. Strober indicate also that spleen cells from humorally tolerant mice also reconstituted lethally irradiated B10. D2 mice to produce cytotoxic (a n d hemagglutinating) antibody in response to antigen. This response resembles closely those of Howard i n studies of tolerance to type 3 pneumococcal polysaccharide (SIII); there was rapid loss of tolerance a m o n g spleen cells transferred to radiated recipients.

To summarize, it seems that there is some kind of s p l i t tolerance with solubilized antigens. The significance of this is probably obvious; it would have some important applications in attempts to manipulate the immune system for organ transplantation and immunotherapy i n malignant diseases.

KAHAN: Will you tell us about the complete tolerance in neonatals w i t h soluble antigen? You are the only group to have demonstrated tolerance.

APELLA: What specifically?

KAHAN: If neonatals were treated with soluble antigens for an extended period of time, do they become tolerant for tissue grafts, so that you have complete tolerance rather than split tolerance.

APELLA: What I know is really very preliminary. The strain combination employed in our studies in B10. A - B10. D2. If the reverse combination were used, B10. D2 - B10. A, about 80% of B10. A recipients of 5-7 x 10^6 (B10. A x B10. D2)F1 bone marrow cells showed complete allograft tolerance to skin graft in contrast to less than 25% in the former combination (B10. A - B10. D2). When we utilized the soluble antigen, it appeared that now we can get in the B10. D2 - B10. A combination skin graft and humoral tolerance in nearly 20% of the B10. A recipients. This combination o f strains differs by one strong alloantigen specificity, 31, so this m e a n s that some other factors besides the type of antigen molecule solubilized are involved, but this is something that I cannot really speculate about.

Table 14

Cytotoxic Antibodies Against H-2^a Antigens

	Thymectomy and reconstitution	
T (intact)	3.6 (8)	
	}	1.2 ± 1.1 (NS)
T (tolerant)	2.4 (14)	
LN (intact)	4.6 (8)	
	}	3.0 ± 1.3 (P = 0.01)
LN (tolerant)	1.6 (11)	
	Irradiation and reconstitution	
BM (intact)	3.7 (9)	
BM (tolerant)	3.3 (7)	

KAHAN: Can we have some general discussion of any points?

McCONNELL: I have the advantage of understanding about five percent of what you people are saying.

(Laughter)

McCONNELL: But we are doing two types of very simple experiments that might possibly have some bearing on the discussion of this afternoon. In some experiments, we took specific antigens into model membranes and have controlled the lateral distribution by manipulating the lipid composition. In such cases, it appears that there is a strong dependence of, for example complement fixation, on this lateral distribution.

The other point that I would like to raise is the question as to whether or not a soluble antigen inhibits the cytotoxic effect of complement. Joe Humphrey and I have seen instances where antigen in apparently low amounts augments the immune lysis of sensitized red blood cells. So it is possible that the effects of solubilized antigens on complement is independent of their immunological specifcity. That is the question.

KAHAN: Actually, that is an exceedingly important issue for the assay, because if we were measuring some non-specific effect on the complement system, the immunological significance of these materials would not be very great.

FERRONE: I don't think there is any evidence which indicates that soluble antigen can absorb to the target cell, and thereby increase lysis by complement and antibody.   What might be possible is just that soluble antigen can affect the target cells in a way which might modify the receptiveness of target cells for antibody, for instance, by the presence of enzymatic activities in the soluble antigen.   What does happen is primarily the opposite effect, that soluble HL-A of H-2 antigens inhibit the activity of the complement system.   We don't know why this happens, but it is another source of the variability that everybody has to face when controlling the inhibition tests.

CUNNINGHAM: I would like to ask both Dr. Appella and Dr. McConnell if their antigens contain carbohydrate.

McCONNELL: Ours certainly did not.   It was a protein derived from red blood cells.   When the protein was present in solution in large amounts, it inhibited complement mediated lysis.   So in effect, it is biphasic.   When we reduced the amount of soluble antigen, it actually augmented complement mediated lysis of cells.

The mechanism that we suspect is involved is a critical number of antigen molecules must be in close proximity to the target to start the complement events moving. It looks as though there is a cross-linking between the antigens on the surface and the antigens free in solution, which can provide that critical number necessary to start the reaction.

KAHAN: Dr. Apella, do you have carbohydrate?

APPELLA: Yes. In the H-2, there is a large chain and a beta-2 microglobulin. We do not know if carbohydrates are distributed between the two chains. In the HL-A, it is quite clear that carbohydrates are on the large component.

CUNNINGHAM: Your purified component has carbohydrate?

APPELLA: Yes.

KAHAN: Do you have some information, Dr. Cunningham?

CUNNINGHAM: No, I just remember that Ralph had some comments that there is no carbohydrate on the HL-A and I just wondered.

KAHAN: No, that is not true, the papain preparations are known to carry carbohydrate, while those derived by other methods apparently do not contain carbohydrate.

REISFELD: I know there is by now a seven-year old controversy about carbohydrate on HL-A molecules. All I can say is, that whether one detects carbohydrate depends somewhat on the method of antigen solubilization applied. Second, it has, by the way, been shown very clearly by Nathanson, that the carbohydrates present on H-2 antigen have no detectable effect on the H-2 antigenic activity.

PINCUS: I think one of the problems that Barry and Ralph alluded to is the difference in solubilization. I would like to say in some preliminary results we have been looking at what is happening to the intact cell before any components are solubilized. Platelets which contain HL-A antigens have three glycoproteins. The glycoproteins in the platelet are of higher molecular weight unlike the other nucleated cells, the smallest being 90-100,000. This is larger than the reported molecular weight of HL-A. It has been shown by Philips who has iodination of the platelets, that trypsin cleaves all the iodinatable portions of the glycoprotein. We also know from work of Jamieson that 50 percent of the platelet sialic acid can be removed by trypsin. Finally, Phillips has also shown that trypsin only cleaves the

three glycoproteins and not the four other iodinatable proteins. In collaboration with K. K. Mittal, we have treated platelets with trypsin, and
determined whether this alters their ability to inhibit cytotoxicity. As we
have found no significant differences in inhibitory activity of platelets
treated with trypsin for 30 minutes, this would seem to suggest that the
HL-A antigen, untouched by any isolation method are probably not glycoproteins.

CUNNINGHAM: I would like to straighten out the confusion. I still do not
know whether or not the HL-A antigens contain carbohydrates.

KAHAN: One runs into the dilemma of knowing what the antigenic composition on the cell surface as opposed to what is present in the solubilized
derivative. On the membrane, I think there is little doubt that there is a
carbohydrate moiety associated very closely with the polypeptide determining H-2 and HL-A. When you use papain to extract the surface, you
get a material containing carbohydrate and protein. When you use other
techniques such as sonication and 3M KCl, the moieties with HL-A activity do not contain carbohydrate at the one percent detection level. The overwhelming evidence is the allotypy is related to protein. There is no evidence for allotypic carbohydrate determinants.

REISFELD: Just let me mention that we found in preliminary experiment
that HL-A antigens in human serum have no detectable carbohydrates once
they have been thoroughly purified. This was determined by gas chromo-
lographic analysis performed by Dr. Dale Sevier in my laboratory.

KAHAN: We will take a short break, and come back to talk about other
systems in which the induction of immune unresponsiveness may be related to the activity of soluble cell surface antigens.

KAHAN: Next, we would like to focus on tumor immunology and specifically to take up where we left off earlier, namely the physiology of the
immunological unresponsiveness in the tumor-bearing host. Dr. McKhann,
is there information that the unresponsive state may be related to excess
circulating antigen, thereby being a natural case of the artificial transplantation model created by Dr. Appella and Dr. Law.

McKHANN: 1. Blocking Factors in Serum: The initial demonstration
that serum factors could "enhance" tumor growth in vivo was by Kaliss.
Although circulating antigen was not eliminated from consideration, most
studies concluded that antibody was the mediator of tumor enhancement.
It is widely speculated, but by no means proven, that the in vitro

counterpart of enhancement is the capacity of serum factors to block cellular cytotoxicity. The earlier studies of the Hellstroms again indicated that the blocking material was specific antibody. However, with Sjogren they subsequently obtained evidence that antigen-antibody complexes were the blocking factor. More recently, a study of Baldwin has provided evidence that free circulating antigen is able to combine with immune lymphoid cells and specifically inhibit their capacity to attack target cells. At the present time, the interrelationship between these three materials, antibody, antigen-antibody complex, and free antigen, in the circulation of tumor-bearing individuals is still undergoing clarification. While each may in some way interfere with cellular immunity, it is certain that circulating antigen is capable of suppressing the tumor-specific activity of lymphocytes.

II. Suppression of Lymphocyte Responsiveness By Soluble Tumor Antigens: Our own laboratory has been concerned with the immunologic effects of soluble tumor antigen, particularly as it may be present in the circulation of tumor-bearing individuals. A model was developed using transplanted methylcholanthrene-induced sarcomas of mice to evaluate the effect of tumor growth on the capacity of lymph node cells to respond to stimulation in vitro upon exposure to intact cells or soluble antigens of the same tumor. It was found that lymph node cells from normal mice, or from those preimmunized against or bearing small transplantable tumors were able to undergo significant stimulation in vitro upon cocultivation with cells of the same tumor. The assay was similar to mixed lymphocyte cultures except that the tumor cells, blocked with mitomycin-C, were the stimulator cells and lymph node cells were the responders. Maximum stimulation occurred on the 5th and 6th days of exposure.

When the responding lymph node cells were recovered from animals bearing tumors, it was noted that those from animals with small tumors responded well to stimulation by the same tumor cells in culture while those from animals bearing large tumors did not. The transition point was about 14 days, by which time the tumor had reached the size of one centimeter in diameter. Prior to that time, there was a progressive increase in the background uptake of tritiated thymidine by lymph node cells from tumor-bearing animals that were grown alone in culture, receiving no further stimulation. In contrast, there was a very dramatic increase in the capacity of the same cells to undergo stimulation when exposed to tumor cells in culture. By ten days, the background uptake of thymidine by unstimulated cells reached a plateau that was three times the uptake of lymph node cells from normal non-tumor bearing mice but the cells responded well to stimulation in vitro. However, abruptly by 14 days, the capacity of the cells to undergo further stimulation in vitro dropped to much lower levels. If the growing tumor was removed, within about one week, the lymph node cells regained their capacity to respond to stimulation in vitro. This was associated with a return of their background activity to normal levels. These findings indicate that the lymph node cells underwent maximal stimulation in vivo for about 14 days after which the tumors appeared to overcome the capacity of the lymphoid cells to undergo further stimulation. The cells that had undergone such stimulation in vivo were essentially refractory to further stimulation and the pool of responsive cells was exhausted in 14 days. The return to normal activity following removal of the tumor probably presented repopulation with new, unstimulated cells.

Lymph node cells used in these studies included not only those close to the tumor but also those remote from it, indicating that the stimulation and refractoriness was expressed throughout the body. This suggested that it may be mediated by circulating tumor antigen produced in sufficient quantities to saturate the immune system at about 14 days. To evaluate this, soluble tumor antigens were prepared from the same tumors. After demonstrating the antigenicity of these preparations by inhibition of cytotoxic antibody (as measured by H^3 thymidine uptake by tumor cells), the soluble antigen preparations were tested for their effect on lymphocyte stimulation in vitro. Active preparations were able to stimulate normal, non-immune lymph node cells in 5-6 days and to stimulate lymph node cells from immune mice at about 48 hours. Following stimulation with soluble antigen, these same lymph node cells were refractory to further stimulation in vitro by cells of the same tumor for 5 or more days, as long as the cultures lasted. Serum from mice bearing large tumors similarly showed a capacity to stimulate lymphocytes in vitro, again inducing a refractory state in these lymphocytes to further stimulation with intact tumor cells. These same sera showed evidence of soluble antigen in that they inhibited antibody mediated cytotoxicity. Serum from animals bearing smaller tumors did not show evidence of circulating antigen by either of these two methods. These findings indicate that the lymph node cells were stimulated by circulating tumor antigen in vivo to a point where they essentially became refractory to further stimulation. This stimulation and refractoriness could be duplicated in vitro using soluble tumor antigen or using serum from animals bearing large tumors. It appears that the immune response was saturated and essentially overcome by excessive production of antigen by about 14 days following which free antigen could be found in the circulation. The protective value of this is shown in Figures 55 and 56.

Of importance in this consideration is the question of whether the immune response is maintained at a high level in the presence of large amounts of antigen or whether it is centrally inhibited with the development of "partial tolerance" of that particular antigen (Figure 57).

IMMUNE SURVEILLANCE AND ESCAPE

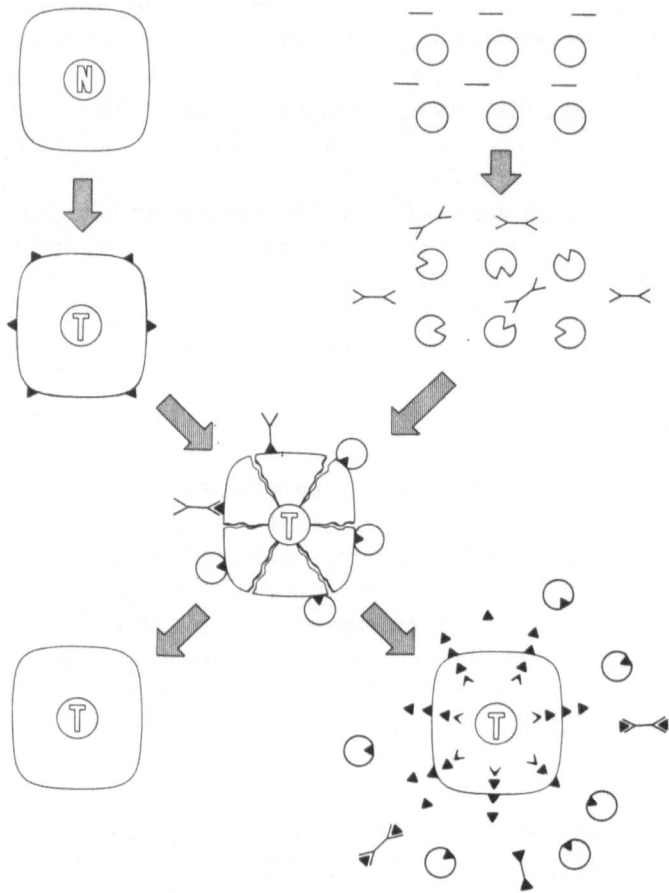

Fig. 55 . In becoming malignant, the normal cell acquires surface antigen that activate the immune response. To escape, the tumor may rid itself of these antigens while remaining malignant or it may go into production of antigen for "export", flooding the system with circulating antigen.

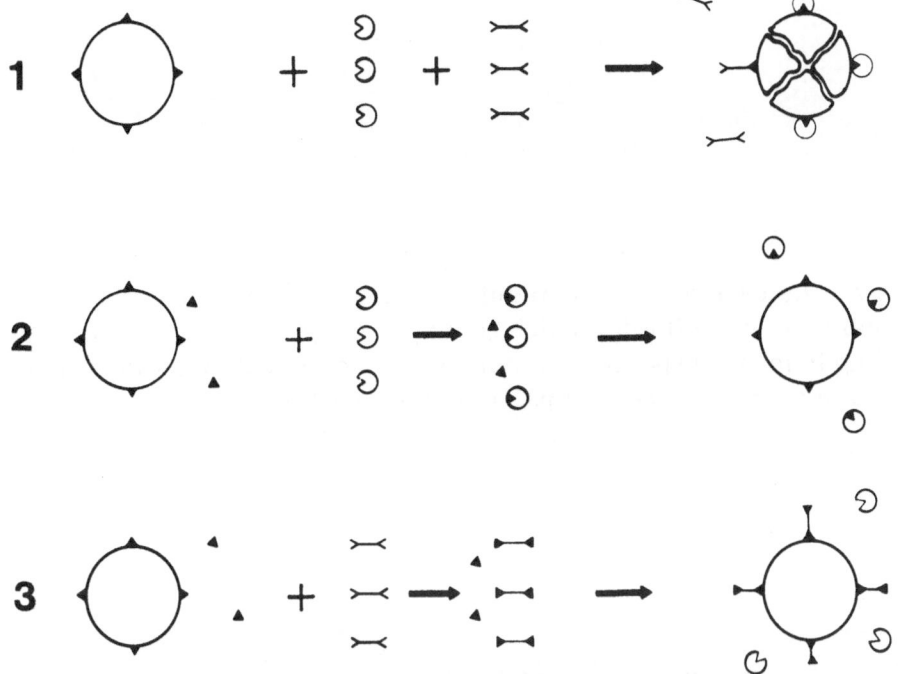

Fig. 56. Target cell killing by immune cells or antibody (1) can be blocked or inhibited by release of circulating antigen by the tumor (2) and (3).

Of importance in this consideration is the question of whether the immune response is maintained at a high level in the presence of large amounts of antigen or whether it is centrally inhibited with the development of "partial tolerance" of that particular antigen (Figure 57). It is important to differentiate between these two possibilities because the therapeutic routes to circumvent them may be entirely different. Maintenance of high levels of production of antibody and immune cells in the presence of excessive antigen would require attack directly on the source of antigen or on the mechanism of its disposal. On the other hand, central inhibition of the immune response may require steps similar to those used for breaking partial tolerance including the introduction of non-tolerant or specifically immune lymphoid cells or further immunization with slightly altered antigens.

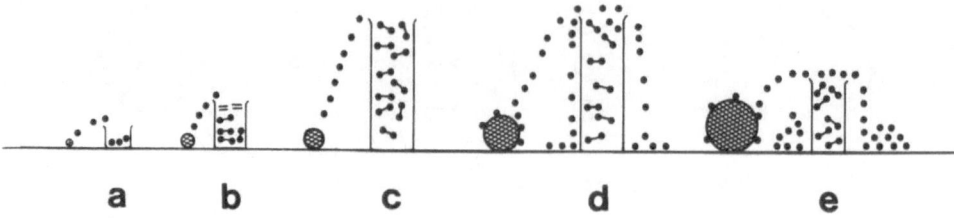

a   b   c   d   e

Fig. 57. Release of antigen stimulates the specific immune response maximally (a - c), following which free antigen appears in the circulation (d). The immune response may remain maximally active (d) or it may be reduced in activity, becoming partially tolerant (e).

III. Tolerance and Autoimmunity: Burnet and Fenner originally developed the concept of tolerance by which the individual must somehow avoid immune attack against his own "self" antigens. The proposed mechanism of this was "clonal deletion". Each individual is endowed with genetically specified clones of cells enabling him to respond to a wide but not unlimited variety of antigens, including his own self antigens. However, in order to avoid immunologic self destruction, those clones that are directed against self are deleted or suppressed, ordinarily never to function again. Against this background, three mechanisms have been proposed for the several known autoimmune diseases. The first of these is that by some mutation, or combination of mutations, a forbidden clone comes back into existence and actively expresses itself against normal tissue antigens. Alternatively, the cells or antigens responsible for autoimmune disease may normally be hidden from the immune response. Such antigens would have clones of lymphoid cells capable of responding to them but would normally find safety in being isolated from the immune response. This would be the case with internal cellular antigens or with cells in privileged locations such as the meninges, the anterior chamber of the eye, the prostate gland, etc. Breaking down these barriers would expose the immune response to normal "self" antigens that it was not intended to see. Finally, exposed antigens could undergo subtle changes that allowed them to stimulate clones of immune cells not included in the original deletion or suppression. These antigenic changes could be brought about by virus infections, cellular toxins, drugs, or malignancy.

An alternative concept of tolerance and autoimmunity rests more heavily on the continued presence of circulating antigen. This proposes that normal cells release large quantities of normal antigens into the circulation, saturating the immune response for that particular antigen. Under these circumstances, there are no forbidden clones and the individual is continually immunized against himself. He escapes from this immunization by continuously inhibiting the immune response against himself with overwhelming amounts of self antigen. Autoimmunity would then occur under circumstances where any particular self antigen, presumably a tissue-specific antigen, was no longer released in sufficient quantities to saturate the immune response to that antigen, allowing the immune response to attack the cell or tissue directly. This could occur in the presence of (a) abnormal growth in which all material were used for the production of new cells and none released into the circulation, (b) loss of cell mass below critical levels, (c) decreased production of surface antigen materials, (d) decreased release of antigens from the cell, or (e) breakdown or autolysis of antigen.

The evidence for this mechanism is as follows: 1) induced tolerance requires the continuous presence of the tolerogenic antigen. When tolerance is allowed to lapse, it is replaced by immunity and not by immunologic "neutrality". 2) Serum of tumor-bearing individuals contain antigens that inhibit cellular cytotoxicity to tumors. This appears to be related to the size of the tumor. 3) Fibroblasts grown in tissue culture can be killed by autologous lymphoid cells if lymphoid cells are maintained in culture for several days. This can be inhibited by the presence of autologous serum, either during preservation of the lymphocytes or at the time of their exposure to the fibroblasts. The implication of this is that lymphocytes acquire the capacity to react against the autologous cells after they have had an opportunity to dispose of inhibitory antigen either by pinocytosis or by extrusion into the medium. Continuous exposure or re-exposure to the antigen maintains the blockade on the lymphocytes. 4) Plasma has now been shown to be a very rich source of normal HL-A antigens indicating that there is a great deal of this material in the circulation.

IV. Non-Specific Immunosuppression in Malignancy: It was originally noticed in patients of Hodgkin's Disease and then in others with a variety of advanced malignancies that the immune capacity of the patient was impaired. Techniques are not available to enumerate T and B cells and to measure their capacity to respond to non-specific mitogens. In vivo correlates include skin testing to a variety of antigens. Our own studies have shown that a significant number of patients with relatively early tumors also show depressed immune capacity as measured by these crude parameters. While it is easy to postulate a mechanism for specific immunosuppression, this non-specific loss of immune capacity is more difficult to encompass. However, the possibility again exists that this is related to excessive production and release of tumor-specific antigens in the circulation. Witz has reported that mice injected with tumor extracts show a decreased capacity for plaque and rosette formation to an unrelated antigen, sheep RBC's, suggesting that large amounts of circulating tumor antigen may bring about an immunosuppression that extends beyond that directly related to the rumor.

V. Summary: Soluble tumor antigen liberated into the circulation interacts with the specific immune response by stimulating lymphoid cells, rendering them refractory to further stimulation. When exposed to large amounts of antigen, the same cells are incapable of cytotoxic interaction with the tumor target cells.

Immunologic tolerance may be the result of saturation of an active immune response with large amounts of "self" or exogenous antigens. The immune response may remain active in the face of this or may undergo partial regression. Autoimmunity would result when the release of self antigen falls below the level required to saturate and supress the specific immune response for that tissue's antigen.

Non-specific suppression of the immune response, as seen in patients with a variety of malignancies, may also be the result of release of large amounts of tumor-specific antigen.

KAHAN: One question before we open this for discussion. Charlie, could the active moiety in the circulation be an antigen antibody complex as well as, or instead of, soluble circulating tumor-specific antigen?

McKHANN: It could be antigen antibody complex. We are just now taking that apart. We don't find it usually in the animals before about fourteen days.

PELLIS: Just one question regarding your serum antigen. Have you been able to display specificity in your lymphocyte stimulation assay?

McKHANN: Both the serum antigen and its capacity to stimulate the lymphocytes and the antigen recovered from the tumor cell are specific. Lymphocytes from animals with a different tumor do not undergo stimulation. Cells exposed to muscle extract do not undergo stimulation.

BACH: Do you know whether the percentage of T cells, and the numbers of macrophages are within normal limits in the peripheral blood of the cancer patients that show a decreased response?

McKHANN: Edmund Yunis looked at this and found that patients with colon carcinoma had quite a reduced number of T cells, averaging about 16 percent T cells.

BACH: That might suggest an explanation for your last finding, the decreased response was for a decreased absolute number of responding cells. If the percentage of T lymphocytes was lower, it would explain your results. As you know, others have shown no decrease in responsiveness in mixed culture in cancer patients.

KAHAN: I would like to bring in a piece of data from the situation of clinical transplantation, and ask Charlie whether it is consistent with his observations in the tumor situation. A system developed by Dr. Baldwin Tom in our laboratory measures the reactivity of peripheral white blood cells derived from patients at intervals after transplantation against kidney monolayers grown in tissue cultures. In the absence of immunocompetence, the white cells show, if anything, a diffuse attachment to the monolayer. With the acquisition of immunocompetence, the pattern of aggregation shows rosettes. This phenomenon is called the leukocyte aggregation assay. The pattern of reactivity is specific for the determinants SD. However, since SD and LD determinants in man show marked linkage disequilibrium, the LD genes may also be reflected by the specificity. Thus, in the case of a transplant yielding sensitization against antigen HLA-2 (the rest of the HL-A phenotype of donor and host being the same), the reactivity is solely toward HLA-2 and the cross reactive antigen W28. There is no reactivity against other antigens.

Confusion arose when the assays were performed only with peripheral blood leukocytes. For example, one young man who had cytotoxic antibody against the lymphocytes of approximately 85 percent of the population (which means he was extensively pre-sensitized) displayed massive aggregation against donor cells the sixth day post transplant. The young fellow was doing just fine and he continued to do well. Only some weeks later did we think to add his serum to the leukocyte reaction. Indeed, there was blocking activity which completely precluded the expression of the cell mediated activity.

Activity of the lymphocytes which could be demonstrated when they were reacted alone, was completely aborted in the presence of serum. Dr. Tom incorporated this into the routine assessment, and it seemed that the effects of serum are very important in pre-sensitized patients for one thing. The effects are frequently more important early, that is, in the immediate post transplant period, at a stage when you would expect the graft to be releasing a lot of antigen as compared to later when one would expect relatively little circulating antigen.

In one case of an allotransplant engrafted into a thirty year old woman, there was a nice fall in serum creatinine and rise in the excretion of radiohippuran. In the first few weeks, the activity of her peripheral leukocyte was blocked by her serum. At about forty days, there was a loss of this pattern, and an emergence of a synergism between serum and cells. Thereafter, there occurred a clinical rejection crisis. So the question that arises from Charlie's comments is whether this phenomenon reflects the influence of large doses of the circulating antigen or antigen-antibody complexes on the immune activity of the lymphocytes of the kidney allograft recipient.

McKHANN: I think it certainly could be. We don't know what the timing of release of antigen from a tumor or a normal transplant kidney transplant really is. We can assume, probably correctly, that the rate or amount of antigen released by a tumor goes up with the increasing size of the tumor. Dr. Yarlott, in my laboratory, is trying to manipulate cell populations in vitro. He has been absorbing into monolayers of tumor cells, lymph node cells from appropriate mice. If the donors are normal mice, the number of cells taken out by the monolayer is not very great.

However, if the lymphoid cells come from tumor bearing mice, a very large number of cells are taken out. It is a much larger number than one would expect on the basis of straightforward specific immunization. We are seeing recruitment or non-specific stickiness, is not known yet, but there is no question that the aggregation that you are seeing of lymphocytes around the target cells can be not only counted and measured, but probably can be utilized.

SUNDHARADAS: Have you tried to use antibody against the tumor antigen to treat patients and see if that will help?

McKHANN: I am one of the members of the old school. Prior to today -- when we heard about using antibody prepared with soluble antigen, I felt that the use of xenogeneic antibody prepared against intact cells was pretty disasterous.

SUNDHARADAS: Since you are saying that there is a flooding of tumor antigen in the system and that may help to protect the tumor from the immune system, this maybe a way to suppress the tumor antigen in the system.

McKHANN: The problem is, if you bind the antigen circulating in the system to a really foreign protein, you may create more trouble than you want to have.

KAHAN: I wonder if Neal Pellis could tell us what evidence there is from in vivo studies that large amounts of exogenously administered

antigen interferes with the immune response. One of the most conventional ways of examining this question is in terms of immunogenicity, the
capacity of the extract to immunize the host so he can reject a syngeneic
tumor to which he is normally susceptible.

PELLIS: There is a little doubt that on the surface of m a n y neoplastic
cells are antigenic determinants which differentiate them from their normal counterparts. Many investigators have shown that these markers can
serve as targets for host immune responses. Since it has been shown that
responses to t h e s e determinants can be destructive to malignant c e l l s,
tumor specific antigens are a logical focus for immunotherapy.

We have been studying tumor specific antigens in s o l u b l e
form and evaluating the preparations on the basis of their in vivo immunogenicity. Testing preparations by other methods (specific lymphocyte
performances) may reflect the quantity of soluble tumor specific determinants, but it cannot assess the immunoprotective potential of s o l u b l e
antigens.

The animal model chosen for these studies was the chemically induced murine sarcoma system in C3H/HeJ female mice. T h i s
model was selected for several reasons: 1) the histocompatibility determinants of mice have been well elucidated; 2) chemically induced tumors
possess marked individuality; and 3) a large body of information has been
accumulated concerning the biology of these tumor specific antigens.

There are numerous methods available for solubilization of
cell surface moieties. Several methods which have been used to solubilize histocompatibility antigens are known to release TSTA from a variety
of tumor cells. Holmes, Kahan and Morton showed that low frequency sonication of tumor cell suspensions affords soluble TSTA which possess immunogenic activity. Oettgen and co-workers solubilized TSTA from guinea
pig fibrosarcomas using hypotonic salt, and were able to immunize hosts
to retard the growth of a tumor challenge. Rapp's group solubilized TSTA
from guinea pig hepatomas. These materials elicited immune performances such as macrophage migration inhibition, delayed cutaneous reactivity,
antigen specific lymphocyte proliferation, and second-set tumor graft rejection.

In our experiments, sarcomas were induced in C3H/HeJ female mice by subcutaneous administration of 3-methylcholanthrene i n
trioctanoin. Ninety to 120 days later, subcutaneous nodules became pal-
pable. Mice with tumors were sacrificed and single cell suspensions were
prepared by trypsinization. The minimum tumor dose (MTD) of the transplantable neoplasms was determined by the method of Gross. Two tumor
lines, MCA-C and MCA-F had MTD's of approximately 10^2 in their fourth
and fifth transplant generation.

The extraction of TSTA from these tumors was done u s i n g

the 3M KCl method of Reisfeld and Kahan. Soluble TSTA was defined as material resistant to sedimentation of 164,800 g. The soluble fraction of the KCl extracts was concentrated by dialysis against 50% sucrose solution and then re-equilibriated to isotonicity by saline dialysis. Insoluble nucleoprotein was sedimented from the dialysan by centrifugation at 48,000 g. The supernate of this centrifugation was administered subcutaneously to syngeneic mice.

Ten days post immunization, the animals were challenged subcutaneously with the immunizing tumor on one side and an indifferent tumor on the contralateral side.

When mice immunized with 1.63 mg of crude MCA-C extract were challenged with 10^5 MCA-C and 10^5 MCA-F, (Fig. 58) the incidence of MCA-F was about equal in the immunized and the untreated groups.

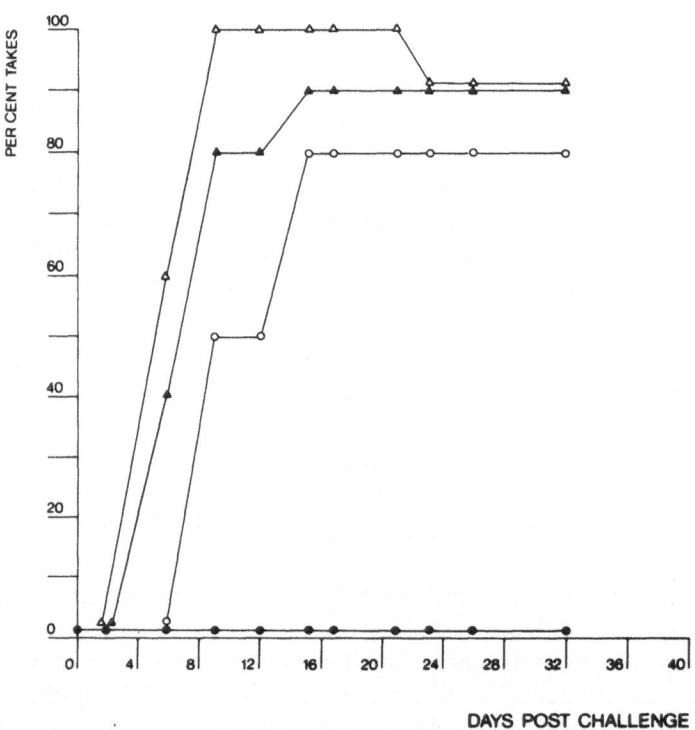

Fig. 58. In vivo immunogenicity of soluble TSTA prepared from MCA-C. Mice were immunized with 1.63 mg of soluble MCA-C and then challenged 10 days later with 10^5 MCA-C and 10^5 MCA-F.   Results are presented as the percent of immunized mice bearing tumors MCA-C (•) and MCA-F (▲) and the percent control mice bearing MCA-C (o) and MCA-F (△).

On the other hand, while the MCA-C challenge proliferated readily in the control group, the immunized group effectively rejected the same challenge. No evidence of MCA-C proliferation was noted in the immunized group up to day 32, when some mice died because of the burden of the indiffered tumor MCA-F. Five survivors of the immunized group had no evidence of MCA-C growth at 65 days, while carrying an MCA-F tumor measuring 2 cm in diameter.

Similar experiments (Fig. 59) were performed with an extract of MCA-F (1.0mg of MCA-F).

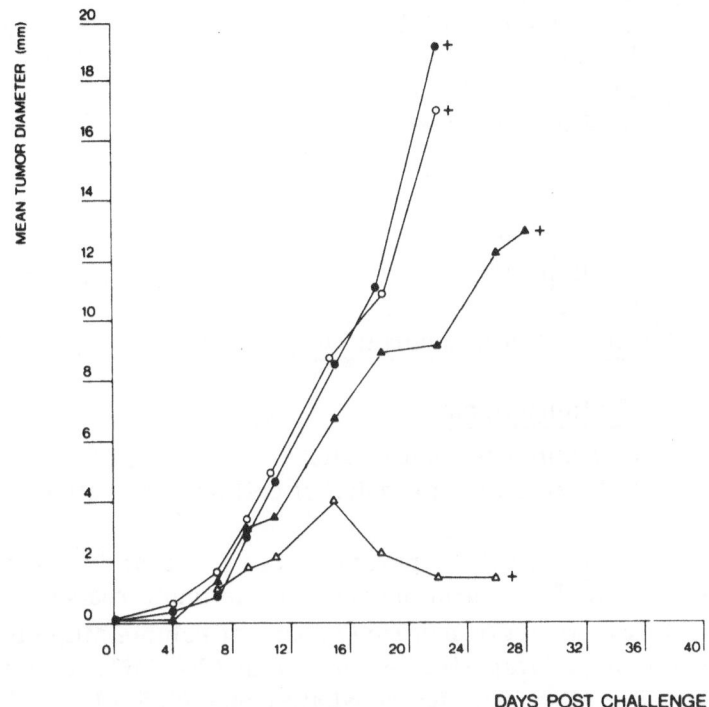

Fig. 59. In vivo immunogenicity of soluble TSTA prepared from MCA-F. Mice were immunized with 1.00 mg of soluble MCA-F and then challenged 10 days later with 10^5 MCA-C and 10^5 MCA-F. Results are presented as growth of tumor (mm) in immunized mice challenged with MCA-C (●) and MCA-F (▲) and in controls challenged with MCA-C (o) and MCA-F (△).

Although this soluble preparation did not afford absolute protection as seen with MCA-C extract, immunized mice displayed retarded tumor growth proliferation followed by total regression. These results indicated that the TSTA of murine sarcomas could be extracted by 3M KCl and that the soluble TSTA induced specific immunity.

The parameters for evaluating the immunogenicity of soluble TSTA are shown below:

1. Nature of the tumor
   a) Solid, blood born, ascites
   b) Minimum tumor dose (MTD)
   c) Growth rate
   d) Transplant generations

2. Host
   a) Histocompatibility
   b) Age
   c) Sex
   d) Health and diet

3. Dose of antigen
   a) Cell equivalent
   b) µg protein

4. Interval of immunization

5. Challenge load
   a) Number of tumor cells
   b) Degree of expression of cell surface antigen

Having controlled the first three parameters, we examined the effect of the dosage of soluble TSTA upon the resultant immune response.

Mice were given various doses of soluble MCA-F antigen and challenged 10 days later with 10^5 MCA-F and 10^5 MCA-C, or approximately 1,000 MTD. (Fig.60). Intermediate doses (0.5 - 1.0 mg) of soluble TSTA possessed immunogenic activities whereas the higher doses did not afford tumor protection. It appears from this data and some preliminary data from extracts of MCA-C that soluble TSTA are immunogenic only over a narrow dose range. The MCA-C challenge exhibited the same growth profile regardless of the administered dose of soluble MCA-F, therefore, specificity was maintained in all doses.

The null effect observed in mice treated with high doses of TSTA may reflect specific immune unresponsiveness, or possibly an enhancement phenomenon. It may be that administration of high doses of TSTA mimic the concomitant immunity observed in hosts bearing established tumors.

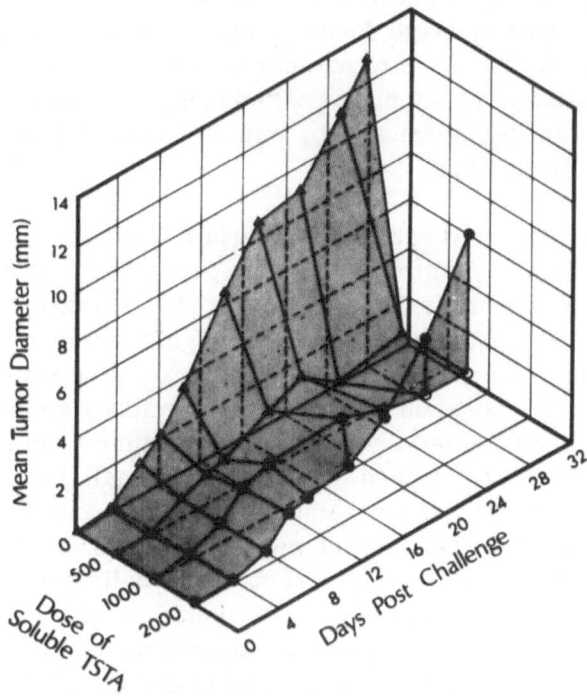

Fig. 60. The effect of immunization with various doses on the growth of MCA-F. Mice immunized with 0, 0.5, 1.0, and 2.0 mg. soluble TSTA were challenged 10 days post immunization MCA-F. Results are presented as man tumor growth (mm).

In conclusion: 1) Tumor specific transplantation antigens of chemically induced murine sarcomas can be solubilized from the cell surface using hypertonic salt solution; 2) Soluble preparations of tumor specific transplantation antigens display in vivo immunogenicity and afford varying degrees of protection against challenge with viable tumor cells; 3) Soluble TSTA are immunogenic over a narrow dose range and therefore probably perform in a similar fashion to the weak histocompatibility antigens.

KAHAN: Certainly, this data presented by Dr. Pellis suggesting decreased immune responsiveness by large doses of tumor antigen, support the concepts proffered by McKhann. Protection can also be obtained by the action of cytotoxic antibody directed toward tumor-specific antigens. Is there any information about the interaction of soluble antigens with cytotoxic antibody against solid tumors?

RISTOW: We have extracted antigenic materials with Shell non-Idet P-40 detergent from a methlcholanthrene sarcoma grown in C57Bl10SN mice and purified them by means of DEAE cellulose chromatography and iso-electric focusing. The protein peaks containing antigen material were identified by means of a sensitive humoral cytotoxicity assay, developed by Dr. Pat Cleveland, based on the capacity of antibody and complement to inhibit the uptake of tritiated thymidine of plated tumor cells in culture. Antiserum for the test was raised in syngeneic animals by giving the animals two tumors in succession, amputating each, and then giving the animals inoculations of 10,000 cells at weekly intervals for 2 to 3 weeks.

Target cells for the assay were sarcoma cells subcultured from 24-hour primary cultures. The cells were plated at 2000 cells per well in microtest 11 plates. Normal immune, absorbed, or antigen inhibited sera were serially diluted in RPMI 1640 containing 30% fetal calf serum and were added to the target cells. After a brief incubation, rabbit complement and tritiated thymidine were added. (Figure 61)

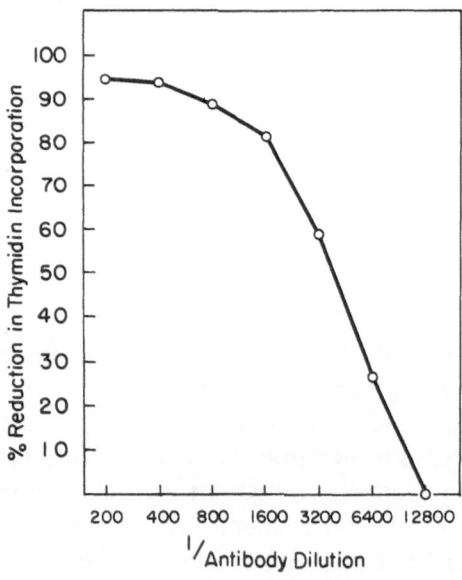

Fig. 61. Titration of MBE antiserum on MBE cells.

A titration of the antiserum made against tumor MBE is shown in Figure 62. Protein solutions were tested for their ability to inhibit the antiserum at dilutions falling at and on either side of the dilution defined as zero cytotoxicity units. The antiserum was absorbed by different numbers of tumor cells, MBE (against which the antiserum was made), and MBK, another tumor of the same series. The results are shown in Figure 62.

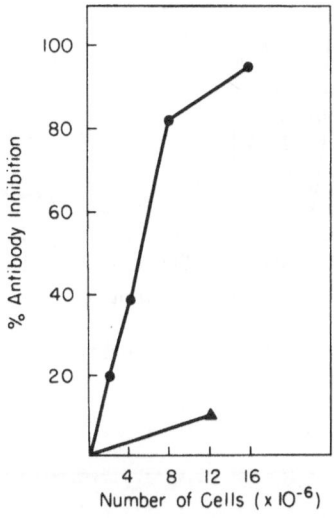

Fig. 62. Absorption of MBE antiserum with varying numbers of MBE and MBK cells. ●——●MBE ▲——▲ MBK)

An NP-40 extract of the MBE tumor was passed through a Sephadex G25 column to remove excess detergent and then was passed down a DEAE cellulose column which was developed by 4 successive buffers of pH 7.4. The peak which contained the most antigenic activity was subjected to isoelectric focusing. The result of this experiment is shown in Figure 63. Peaks 1, 111 and IV were assayed by the humoral cytotoxicity assay. Figure 64 reveals that peak III showed significant activity inhibiting the cytotoxic antibody at less than 0.5 ug/ml.

I think Dr. Reynolds pointed out yesterday, that there are all kinds of difficulties with these detergent extracts of protein. We have also used hypertonic KCl for tumor antigen in our lab and find that they are quite satisfactory. I simply wanted to show an application of the humoral cytotoxicity assay with a detergent extract of antigen.

FERRONE: Is there any reason why you use thymidine to label the cells?

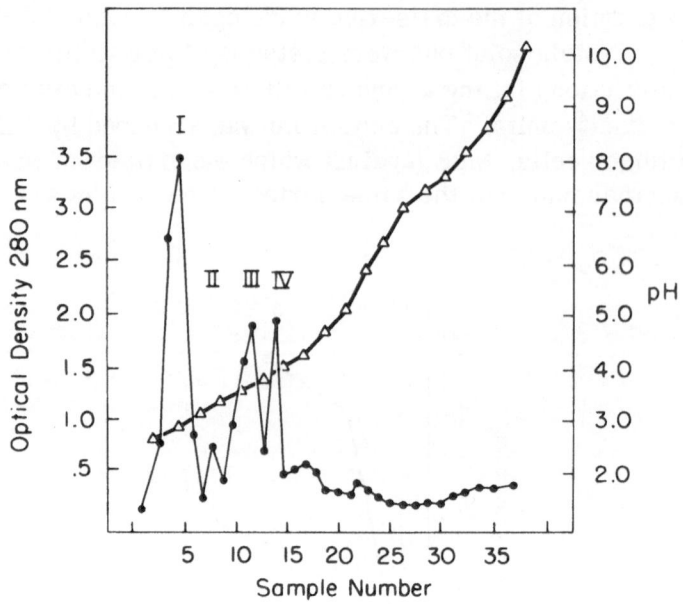

Fig. 63. Isoelectric focusing of antigen containing peak from DEAE cellulose column.

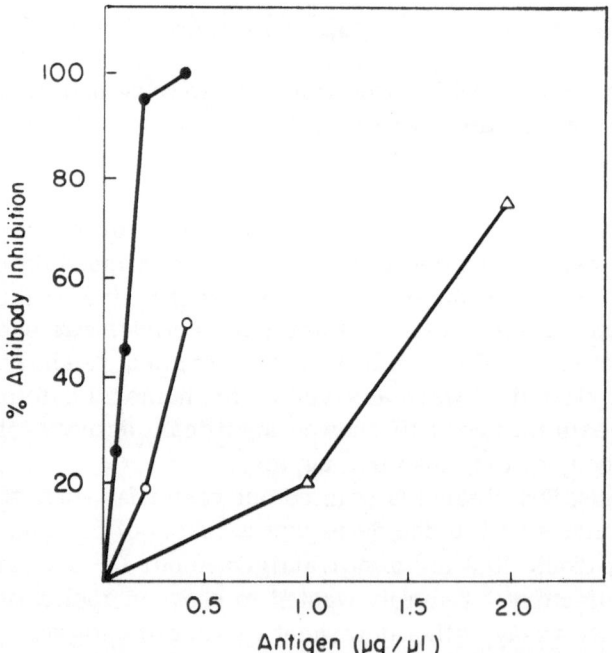

Fig. 64. Assay of peaks 1, III, IV with humoral cytotoxicity assay.
(△ peak 1; ● peak III; o peak IV)

RISTOW: This was the label that Pat Cleveland used when he developed the assay, so I simply continued using it. Alternatively, tritiated uridine may be used in the assay.

FERRONE: Can you use chromium 51?

RISTOW: I have never tried chromium.

KAHAN: My impression of this assay is that it is something different from the cytotoxicity CML, because it reflects an altered capacity of tumor cells to spontaneously divide, as a reflection of cytotoxicity.

McKHANN: This is an uptake assay. The cells are grown in the presence of tritiated thymidine, uridine or even amino acids. We are measuring inhibition of the cells to take up this material during the period of incubation in the presence of antibody and complement. The cells are labeled at the same time they are exposed to the antibody and complement, and we measure decreased uptake.

RANNEY: Do you have controls in which you have not added complement? There are many systems in which you can inhibit thymidine uptake without killing the cells, and this is completely reversible.

McKHANN: All of the controls are included which would be in any cytotoxic assay, including complement alone and antibody alone.

FERRONE: How does it correlate with the conventional cytotoxic assay based on cell lysis? Is it more sensitive?

McKHANN: It is very much more sensitive. It is carried out over a period of 24 hours. Furthermore, most tumor cells will not behave very well with radio-chromium release over that length of time. We think this is a fairly important factor, because chromium release is excellent with lymphoid cells, and with a few solid tumor cells. It has not been very good with sarcoma cells because of the high rate of spontaneous release.

KAHAN: Isn't the cytotoxic assay primarily useful with malignant lymphoid cells?

McKHANN: Chromium release is excellent for lymphoid tumors.

FERRONE: We have adopted the complement dependent cytotoxic test which we have previously utilized for HL-A typing to detect antibodies to melanoma associated antigens. The only difference between the HL-A typing reaction and this melanoma test is that the incubation has to be carried out for 14 hours. For that reason, I was curious to see if your test was less, or more, sensitive than the cytotoxic test.

McKHANN: We compared our assay with chromium release using strong anti-H-2 antisera, and there is essentially no difference. The problem is that the weak antisera against tumor specific antigen takes longer to damage the cell. There is often almost no cytotoxicity at 1 or 2 hours. Unless you can wait much longer, you don't see much cytotoxicity.

KAHAN: Is the presumption that the antibody is of a different quality rather than a different quantity?

McKHANN: I don't think anybody knows.

KAHAN: Any other comments? If not, we might pass on to the last portion of this session which really relates to a quite similar aspect of cell surface function, namely the problem of hormone interaction with plasma membrane structures. I think it was pointed out earlier this afternoon that the immunologist can learn from the endocrinologist, who has the opportunity to use relatively well-defined substances to study surface receptor sites. Our approach to this area will be tripartite. The first point of discussion is hormone binding. In the immunological situations which we have discussed today, no one is using a binding assay for antigen detection. We are just not able to do that today. We don't have even partially purified receptors and we don't have good enough antibody. The second stage after binding may be a translation of the message, a matter that Jack Pincus raised yesterday in terms of adenyl cyclase levels. Translation may be not only by alteration of endogenous plasma membrane adenyl cyclase or other enzyme levels, but also by production or release of factors triggering one target tissue activity as seen with proliferative factors generated by granulocytes. What is the present status of the binding assays of hormone receptors?

KATZEN: The subject of my presentation concerns our studies on the interaction or, more precisely, the reversibility and subsequent inversibility of the interaction, of insulin with the solubilized insulin receptor from rat adipocyte membranes. However, before I discuss our findings, I should first attempt to introduce the relevancy of this hormone receptor to the general subject matter considered at this Conference on "Approaches to the Cell Surface."

       To begin with, the accumulated evidence to date strongly indicates, although admittedly it still remains to be definitively demonstrated, that the receptor site (or sites) responsible for the primary interaction of

insulin with the cell, and that triggers all of the resultant biological effects of the hormone, is located on the surface membrane of the target
cells. In addition, various lines of evidence from our laboratory, derived
particularly from experiments utilizing the lectin Concanavalin A and relevant glycoside ligand probes, clearly favor the view that the insulin receptor is a glycoprotein. Considering the many discussions we have had
the past two days on membrane glycoproteins and other receptors, the significance of this hormone receptor to this Conference becomes apparent.

       In addition to introducing to you the possible implications of our
following findings, the significance of the effects of the interaction of insulin with the solubilized insulin-binding glycoprotein receptor further illustrates the relevancy of this receptor to the topics discussed this morning.
I would like to first outline these implications so that the significance of
the results I will present may be better appreciated. To begin with, under
my main topic entitled "irreversibility of insulin binding induced by soluble
receptor: hormone interaction", I have listed as possible implications:
a) clustering or "capping" of receptors and, b) negative cooperativity of
binding, previously reported by Roth for the intact lymphocyte insulin receptors. Secondly, our findings are relevant to the question of the possible association of the receptor to the enzymatic degradation of the hormone. Finally, they will emphasize the pitfalls resulting from inherent
copurification of "insulinase" (insulin-degrading enzymes) with the receptor. In the latter case, it is easy to see where degrading enzymes that
may interact with insulin would readily contaminate or coexist with receptor preparations if binding to insulin were the only criterion for purification.

       Since little has been said here about hormone receptors, and
since there are probably few endocrinologists attending this Conference,
I would like to begin by giving you a general birds-eye view of the many
intra-cellular and membrane events presumably resulting solely from the
interaction of insulin with the cell surface receptor. In Figure 65, I
simply want to point out these many effects of the hormone (e.g., anti-
lipolytic, lipogenic, glycogenic, anti-gluconeogenic, ionic, transport of
carbohydrates and amino acids, protein biosynthetic, and anti-protein
degradation). While it is believed by some that cyclic-AMP and -GMP as
well as ionic translocation (e.g., via AT Pase) may act as "second messengers" to explain at least some of the effects of insulin, this is still
not generally accepted and certainly not yet proven.

       Figure 66 begins to focus on the thrust of this paper. Here,
solubilization of the membrane binding fraction was accomplished by treatment of adipocyte membrane preparations with the detergent Triton X-100.
The binding component was identified after brief incubation of the soluble
extract with ^{125}I-insulin and passing the resultant solution through a Sephadex G-100 gel filtration column. We have used this as a quantitative assay

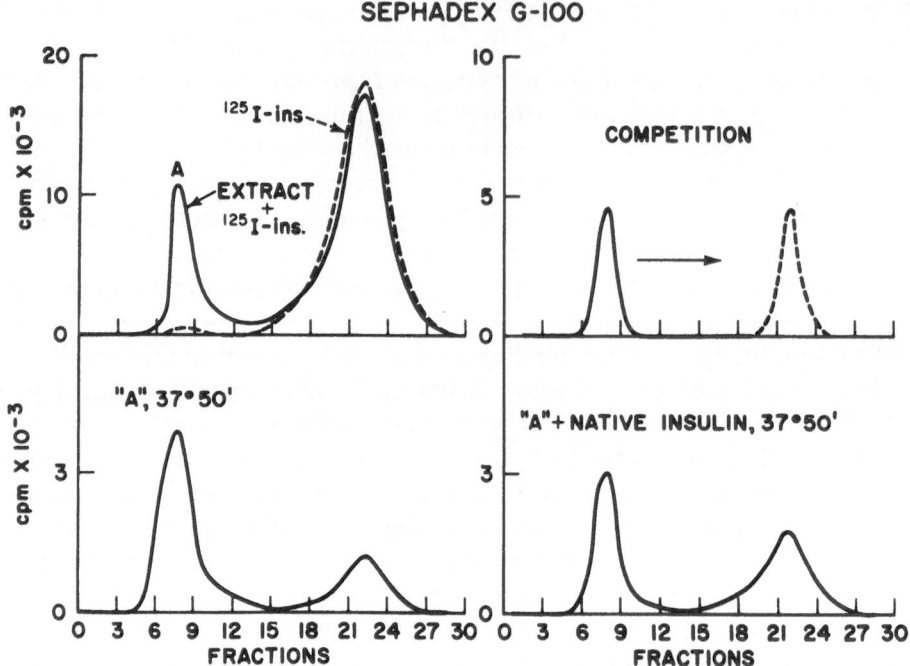

Fig. 65. Sephadex G-100 column separations of bound ^{125}I-insulin from free ^{125}I-insulin and other small molecular weight radioactive fractions.

to measure the receptor wherein the highest molecular weight fraction represents the receptor: ^{125}I-insulin complex appearing in the void volume fractions as a distinct radioactive peak. The excess unbound free ^{125}I-insulin appears retarded (later) in about fractions 20 to 25. Native insulin can readily compete with ^{125}I-insulin for binding to the receptor, thereby completely inhibiting the appearance of the receptor: ^{125}I-insulin peak and placing all of the ^{125}I-insulin in the free ^{125}I-insulin retarded fractions. This is simply a pure competition experiment demonstrating that ^{125}I-insulin appears identical to native insulin in binding to the receptor in this regard.

However, in repeated experiments conducted under many different conditions, we were unable to observe displacement by excess native insulin of the ^{125}I-insulin prebound to this soluble receptor. Thus, while the pooled fractions containing the receptor: ^{125}I-insulin complex retained its elution identity after being passed through a second G-100 column, the complex appeared completely irreversible after it had passed through the initial column. Addition of native insulin did not result in the concomitant appearance of a free ^{125}I-insulin peak.

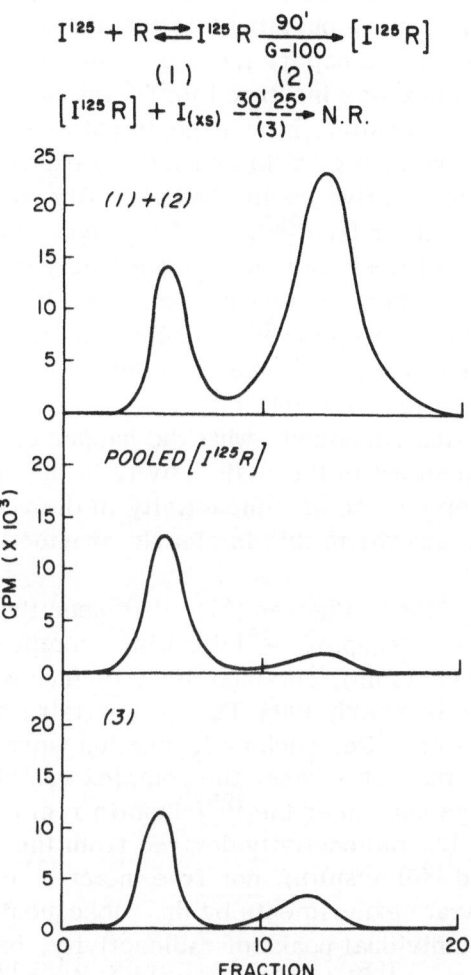

Fig. 66. Same as Figure 65, except including postulated scheme of events.

During the course of these experiments, a paper by Dr. Pedro Cuatrecasas appeared in the Proc. Nat'l. Acad. Sci. in which s h o w e d a complete loss of the high molecular weight receptor: ^{125}I-insulin complex and its replacement with a free retarded radioactive peak after incubation of pooled fractions of the complex for 50 minutes at 37°C with ex - cess native insulin. We were immediately puzzled by this in view of our repeatedly unsuccessful attempt to demonstrate such apparent reversibility. However, what was more puzzling was the absence of a c o n t r o l experience in the Cuatrecasas paper, i.e., no demonstration o f w h a t occurred when the complex was incubated for 50 minutes at 37°C in t h e absence of excess native insulin. In an experiment shown on this slide, it can be seen that comparison with this necessary experiment r e v e a l s that the addition of excess native insulin had virtually no effect o n t h e amount of either the bound or free ^{125}I-insulin peaks. Any loss of radioactivity from the receptor peak was clearly due solely to the time and temperature of incubation. Thus, the Cuatrecasas experiment did not demonstrate reversibility of the receptor: ^{125}-insulin interaction, and his reported Kd of 100 nM insulin becomes questionable. I will return to this point, i.e., the dissociation constant, later.

But the question remained: what did happen during the incubation to cause the replacement of the radioactivity in the high m o l e c u l a r weight peak with the appearance of radioactivity in the retarded region of free ^{125}I-insulin? The answer to this is clearly provided in the next t w o figures.

In the first of these Figures (67), it is seen that while the pooled fractions containing the receptor: ^{125}I-insulin complex and free ^{125}I-insulin (from the G-100 column), obtained prior to incubation at 37° for 50 minutes, were found to be nearly 100% TCA precipitable as expected, t h e retarded fractions obtained after such an incubation were almost completely TCA -soluble. In the latter case, the complex remained 98% TCA precipitable. Thus, the peak near the ^{125}I-insulin region off the column was actually degraded 125 radioactivity derived from the degradation o f insulin (i.e., degraded ^{125}I-insulin), not free intact ^{125}I-insulin. T h i s was confirmed in separate experiments by the subsequent separation and identification of the 3 individual peaks of radioactivity, based upon T C A precipitability of the bound ^{125}I-insulin (complex), free ^{125}I-insulin a n d degraded ^{125}I-insulin.

Furthermore, in the next Figure (68), it is seen that affinity chromatographic insulin - Sepharose columns are capable of simultaneously "fishing-out" both the insulin-degrading enzyme activities and the receptor. Under such conditions, all of the degrading activity, and therefore, t h e degraded ^{125}I-insulin peak seen from the G-100 column, could be elimina - ted.

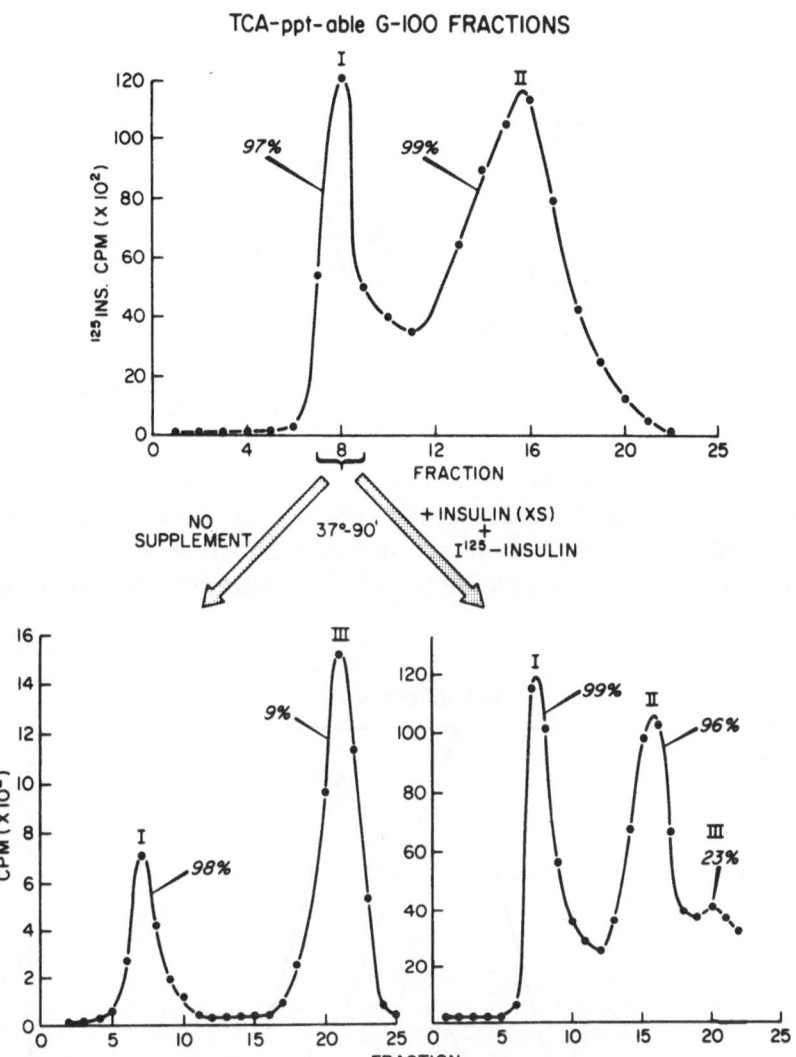

Fig. 67. Same as Figure 65, except degrees of trichloroacetic acid-precipitabilities of resultant pooled peak fractions included. Percents denote percent TCA-precipitable.

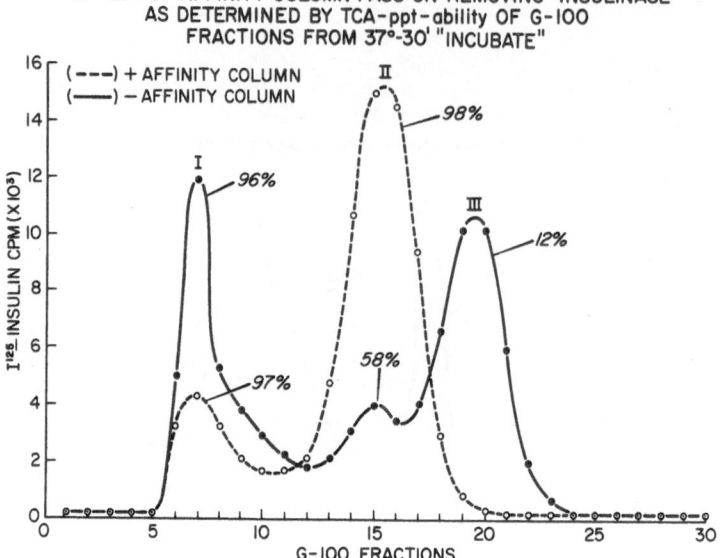

Fig. 68. Identification of bound ^{125}I-insulin, free ^{125}I-insulin and degraded ^{125}I-insulin according to Sephadex G-100 separations after prior passage of soluble extracts through Sepharose and insulin-Sepharose affinity columns, and after incubation of latter eluates with ^{125}I-insulin for 30 min. at 37°C.

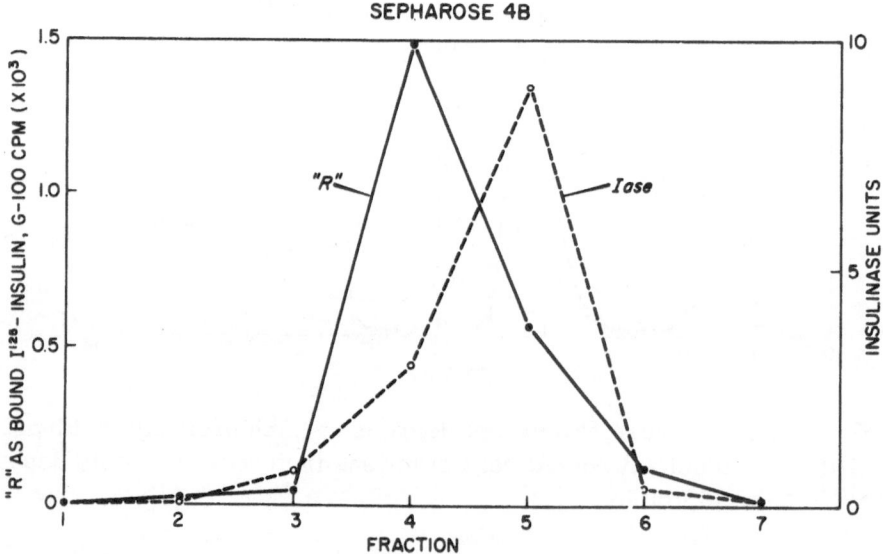

Fig. 69. Partial separation of receptor-binding fraction from "insulinase" degrading activity according to Sepharose gel filtration.

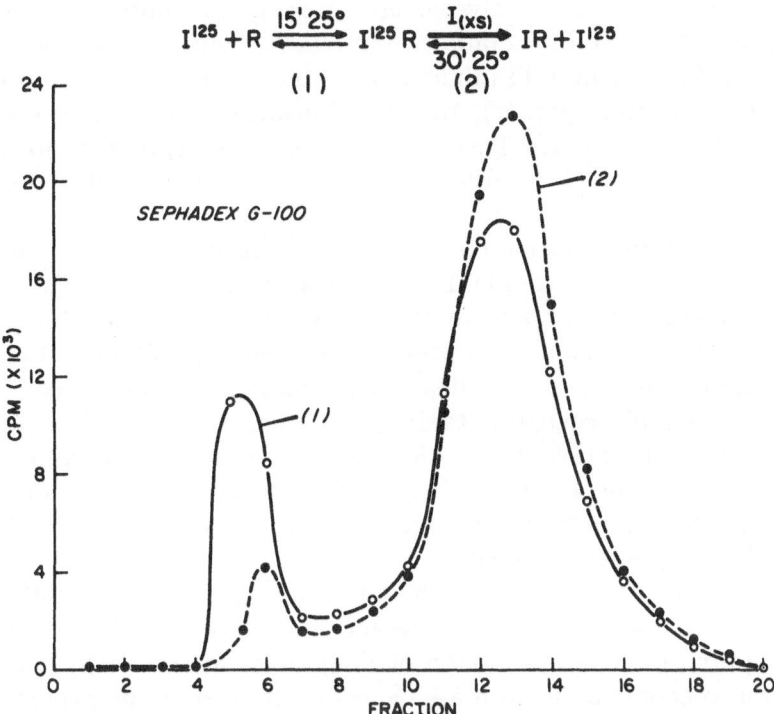

Fig. 70. Displacement of bound ^{125}I-insulin by native insulin (1) a f t e r incubation of ^{125}I-insulin (I^{125}) with receptor extract (R) as determined by Sephadex G-100 column "assay".

Finally, on the next Figure (69), it can be s e e n that conventional methods, as well as a Con A-Sepharose affinity column that o n l y binds the glycoprotein insulin receptor, can effectively separate the "i n - sulinase" activities from the insulin receptor fraction.

This, in the Cuatrecasas' experiments the binding of insulin t o the soluble receptor was not demonstrated to be reversible. The loss of radioactivity from the receptor can now be readily explained by the degra- dation of the ^{125}I-insulin on the receptor and its release as radioactivity in the form of degraded insulin mistakenly identified as intact reversibly de-bound ^{125}I-insulin.

O u r experiments also demonstrate a separation o f receptor from insulinase activity. They also point out the care that must be taken to a v o i d copurification of both binding fractions, and possible mistaken identity of insulinase for receptor.

However, w e have still not answered our initial question o f whether or n o t the soluble receptor:insulin is irreversible. In Figure 70, it can be seen that if excess native insulin is added to a solution containing the receptor: ^{125}I-insulin complex immediately after initial information of the complex (i.e., within 15 minutes after incubation of ^{125}I-insulin with the solubilized receptor extract), the native insulin is shown to displace the ^{125}I-insulin on the receptor. This must be done at 25°C or less to avoid any degradation of the ^{125}I-insulin. Thus, the binding of insulin to the receptor is reversible. Under such conditions, or in true competition experiments, we find a dissociation constant (Kd) of about 1 nMolar insulin, in contrast to the 100 nM reported by Dr. Cuatrecasas.

However, it is important to point out that with time, the receptor:insulin) complex becomes irreversible and even insoluble. Thus, we have been able to observe a time-dependent change in the complex. While we find that the soluble receptor itself appears quite stable and remains soluble with time, its interaction with insulin results in irreversibly-bound hormone and even to insolubility. This may be due to an insulin-induced conformational change in the receptor, or more likely, to aggregation of receptors linked together by insulin.

It may be speculated that the hormone is, therefore, multivalent, as is Con A (which exhibits insulin-like activity at about 10^{-9}M, binds to the soluble receptor, and inhibits insulin binding to the intact cell). The precipitation of the receptor by insulin also appears similar to the precipitation of this receptor and other glycoproteins with Con A.

The irreversibility of the binding of insulin to the soluble receptor in the presence of saturating amounts of insulin may be related to the "negative cooperativity" reported by Roth and co-workers for the binding of insulin to intact lymphocytes. Finally, it may be speculated that aggregation of insulin receptor molecules in solution could have some relevancy to the situation on the intact cell. If so, it may be possible, as is apparently the case for other Con A membrane receptors, that receptors linked to each other by insulin molecules could lead to the type of clustering or "capping" of fluid or mobile receptors on the cell previously discussed at this symposium.

Regarding the latter point, preliminary studies in our laboratory suggest that adipocytes only bind insulin-Sepharose beads (ON) on one surface of the cell, i.e., they only bind one bead per cell, whereas a single bead can readily bind many cells. Thus, it may further be speculated that the interaction of an insulin-Sepharose bead with a cell attracts, or clusters, all of the insulin receptors to the site of interaction of the bead with the cell. In this way, the remainder of the cell surface might lack receptors required for binding further beads.

I throw these speculative matters out to you for your consideration and discussion.

CUATRECASAS: I don't see much disagreement of basic or primary importance but rather very minor issues probably related to different experimental approaches or interpretation. We certainly agree that the receptors are probably glycoprotein, as we described a few years ago, on the basis of enzymic digestions and lectin (soluble and on columns) studies. You quote our experiments on the soluble receptors and reversibility, but not completely accurately. You forget that after heating at 37° with native insulin, the void volume of the subsequent column was <u>still</u> able to bind I^{125}-insulin, indicating that the loss of radioactivity was not simply "degradation" of the receptor. The specific control you referred to as well as this particular one <u>were</u> performed. We are the first to agree, however, that the <u>free</u> insulin, whether just there or whether dissociated from the complex, is rapidly degraded by proteases. The important issue relates to the fact that insulin, while bound to the receptor, is protected from degradation by separate proteases, and that the receptor itself does not have intrinsic degradative capacities related to its function as a receptor. Your data indicating that the Con A-columns <u>separate</u> physically the receptor from the "insulinase" support this quite strongly, and mainly, independent studies also support this. We also don't disagree that with time of incubation of the receptor-insulin complex, the situation with regard to reversibility becomes more complicated and there are an awful lot of artefacts and complications which make interpretation difficult. In sum, I see the differences of opinion so trivial relative to the major findings and to the big issues that unless we are careful to avoid getting lost in <u>apparent</u>, minor discrepancies, we may dissipate our energies so non-constructively that we will lose the forest from the trees and retard new, major breakthroughs.

KATZEN: No control, as you describe, was reported in your publication and we find that ^{125}I-insulin bound to the soluble receptor is not protected from degradation, contrary to your assertion. It remains clear that your "dissociation" must have been due to degradation.

KAHAN: Did you mention capping? Do you think this is really analogous to the capping situation with antibodies. What do you think, Marty?

RAFF: I can't answer that question, but it could be answered by the distribution of the radiolabelled insulin on the fat cells sticking to the beads.

KATZEN: We are tempted to do that.

RAFF: What has been the problem in actually looking at the distribution of insulin on the fat cell. If you can radiolabel it, you can look at the distribution by auto-radiography. Alternatively, you should be able to use labeled anti-insulin antibody.

KATZEN: I think the hang-up has been that of the constant bug-a-boo that has haunted insulin people for years. When one puts iodine on the insulin or other proteins, people have been concerned that it may alter its biological activity and make it irrelevant as a binder. This is one reason that many people have been leary about chemically altering the insulin in any way.

RAFF: I have a suggestion to get at the question of whether X-linking by insulin is required for stimulation. Since tetravalant Con A can induce the same effect as insulin, it would be interesting to know if divalent succinylated Con A can induce the same effect.

KATZEN: I didn't have time to go into our Con A experiments, but for those of you who are not familiar with them, Con A, at about 10^{-9} molar which is about a 50 micrograms per ml., acts in a manner identical to insulin on the adipocyte. It also inhibits insulin binding. The sugars that will knock off Con A from the usual Con A glycoprotein receptors coincide with the glycosides that will dissociate insulin off from its receptor. These studies have been done both in our laboratory with the intact cell receptor, and with the solubilized receptor. There is no question from our results that the Con A can bind to the same binding protein as insulin in much the same fashion as Con A is usually known to bind to ligands, and insulin binds in an apparently identical fashion to its receptor. There is a close correlation between the two.

PINCUS: The fact that you get negative cooperativity in your solubilized preparation is interesting particularly in view of the fact that Rafftery in his work on the cholinergic receptor has found that solubilizing with triton causes the isolated preparation to lose that property. I wonder if you have any information on whether what you are isolating is similar to the aggregated or associated state that you find in the membrane.

KATZEN: I really don't know. We are presently doing studies to see if we can demonstrate a time-dependent increase in the molecular weight of the "complex". We do know it is not a solubility, per se, but I can't go into details now.

SONNENBERG: In an answer to Marty Raff, insulin has been labeled with ferritin and incubated with the adipocyte. It does not show capping in those studies. I don't know the details of those experiments, but they just show general distribution of the ferritin labeled insulin. The latter was biologically active.

We have data with regard to growth hormone, which is consistent with yours, which suggests negative cooperativity. These studies are not with solubilized receptors but with intact hepatocytes. Your studies suggest that there are conformational changes going on within the membrane. The conformational changes we have observed were with extraordinarily small doses. In our experiments, as few as a hundred molecules of growth hormone per cell, or in the case of McConnel and his associates, one molecule per cell have produced uniformal tissual changes. This would suggest that you are getting an overall effect on the membrane and not just a localized effect.

So I would return to your first slide and the slides of others, where the receptor is placed contiguous with the cyclase or the ATPase. It is not necessary to have them juxtaposed that way. The enzymes, as well as the receptors, may be distributed around the whole membrane which is then able to have an overall conformational change and consequent multiple enzyme activation and receptor modification.

KATZEN: All of our studies that I have mentioned here today, have been on the solubilized, dissociated, membrane receptor. I wouldn't stick my neck out and say that our evidence indicates capping. I am just throwing it forth as a bit of speculation and a bit of a possibility.

POLLARD: Just an observation. You report that beads bind to the cells, but that cells don't bind beads.

KATZEN: Most of the time they don't.

POLLARD: If the data of yourself and Jesse Roth and others showing negative cooperative binding is true, then the interaction of the cell with the bead should suppress binding of the other receptors or the cell to insulin. Therefore, your observation would result regardless of what particular mechanism you might want to invoke.

KAHAN: Are there further questions?

McKHANN: Your pictures on the board look very much like antigen:antibody, in "excess" of one or the other. Can you change the ratios and still get the same type of picture? You showed a large number of cells, and rather few beads. Did you try to reduce the number of cells?

KATZEN: Yes, we have reduced the ratio of cells to beads, and we can always show single beads with many cells around each. But there appears to be a much lower frequency of cells with several beads around each, indicating that the cell may have lost its capacity to bind on to more than one bead, i.e., at more than one point, after binding to an insulin-Sepharose bead.

KAHAN: Now, what happens after the hormone is bound. What a r e the immediate consequences of that association.

LIN: For the past f e w years, we* have been studying the hormone r e s - ponse o f t h e adenylate cyclase system. Since adenylate cyclase i s be- lieved to be a component of plasma membrane, hopefully, our approach, with the use of hormone as a tool, will provide further understanding o f the membrane structure and function. I would like to first discuss some of the properties o f the adenylate cyclase system from rat liver and then describe how we attempt to solubilize a n d isolate components of this en- zyme system.

Adenylate cyclase catalyzes the formation of cyclic AMP f r o m substrate ATP. It appears to require sulfhydryl groups for t h e activity. The enzyme in the absence of activator has a substantial level of activity which is known as basal activity. We use a partially purified membrane preparation as the source of enzyme a n d it is activated by glucagon b u t only slightly by epinephrine. NaF has been shown to activate all adenyl- ate cyclase from eucaryotic cells. Recently we have observed t h a t gua- nine nucleotide also activates the e n z y m e at a concentration as l o w a s 10^{-8}M. Now this nucleotide activation has been shown in m a n y other t i s - sues, s u c h a s fat cells, adrenal cortex, erythrocyte, platelet, bladder, thyroid gland, pancreatic islet and brain. We have proposed that th is nu- cleotide effect plays an essential r o l e in the hormone response o f t h e adenylate cyclase system.

Methylene and imido analogs of GTP, Gpp(CH$_2$)p and Gpp (NH)p, which cannot serve as donors for terminal phosphate, activate the enzyme even more markedly than GTP itself. The activation by these two analogs rules out the phosphorylation being involved in the nucleotide effect. Ano- ther analog of GTP, 3'-deoxy-GTP, also activates the enzyme which s u g - gests that cyclic GMP is not a mediator i n t h e activation. None of t h e other guanine nucleotides, including cyclic GMP, GMP, Gpppp and ppGpp, has any effect which reflects the highly specific nature of this interaction. W h i l e pyrimidine nucleotide has no effect, adenine nucleotide activates only at 1000-fold higher in concentration. This extent of activation could be due to the contaminating GTP in the ATP preparation. GDP turns o u t to be a potent competitive inhibitor with about the same affinity for the site.

---

* T h e studies presented here h a v e been performed in collaboration with Y. Salomon and M. Rodbell.

This nucleotide effect seems to interact with glucagon response in an interdependent manner. In the absence of guanine nucleotide, as shown in the panel A of Figure 71, hormone response exhibits lag, i.e., there is a time lag between the addition of glucagon and the onset of its response in terms of cyclic AMP production. The basal activity takes off immediately from zero time in a linear fashion. When glucagon was added at a concentration of $2x10^{-10}$ M, no detectable activation over the basal activity was seen until about 3 minutes after the addition. This lag is gradually shortened with increasing concentration of glucagon as shown in the inset of panel B. When GTP was added, as shown in the panel B of the figure, the lag time was reduced to less than 30 seconds at all concentrations of glucagon tested. The half-maximal concentration for glucagon is reduced from 2 nM in the absence to 0.5 nM in the presence of GTP, thus the adenylate cyclase system becomes more sensitive to hormone response in the presence of nucleotide.

In order to maintain the nucleotide concentration, imido analog of GTP, Gpp(NH), which is resistant to phosphohydrolase activity, was used in the kinetic study of the system. When this nucleotide was added at $5x10^{-8}$ M concentration a lag of about 3 to 4 minutes was observed. This lag was shortened by increasing concentrations or elevated temperature. The addition of 1 nM of glucagon reduced the lag to less than one minute. These results suggest that the binding of each ligand facilitates the transition induced by the binding of the other ligand. The final result of this transition is the activation of the adenylate cyclase. This nucleotide effect persists even if there is no functional receptor as in the case of a mutant cell line of adrenal tumor or when the beta receptor is blocked by propranolol in the frog erythrocyte.

We also know that this membrane preparation binds ^{125}I-glucagon in a slowly reversible manner. As shown in Figure 72, the binding process takes more than 10 minutes to reach equilibrium. If large excess of unlabeled glucagon was added at the equilibrium, there was only a partial reversal of the binding. However, if GTP is also added, the bound glucagon is completely released. When GTP is added at zero time, the binding reaches a reduced equilibrium state much more rapidly. So the guanine nucleotide appears to make the interaction more freely reversible and thus making the enzyme system more responsive to hormone regulation. Since bound glucagon is very slowly dissociated in the absence of nucleotide, one can treat the membrane preparation with various concentrations of glucagon, so to produce various degree of bound glucagon with the excess free glucagon removed. When the adenylate cyclase activity of this treated membrane with bound glucagon was measured, as shown in Figure 73, a hyperbolic relationship between the degree of occupancy and the extent of enzyme activation was observed. The shape of curve approaches linearity when nucleotide concentration in the assay medium is reduced. Therefore,

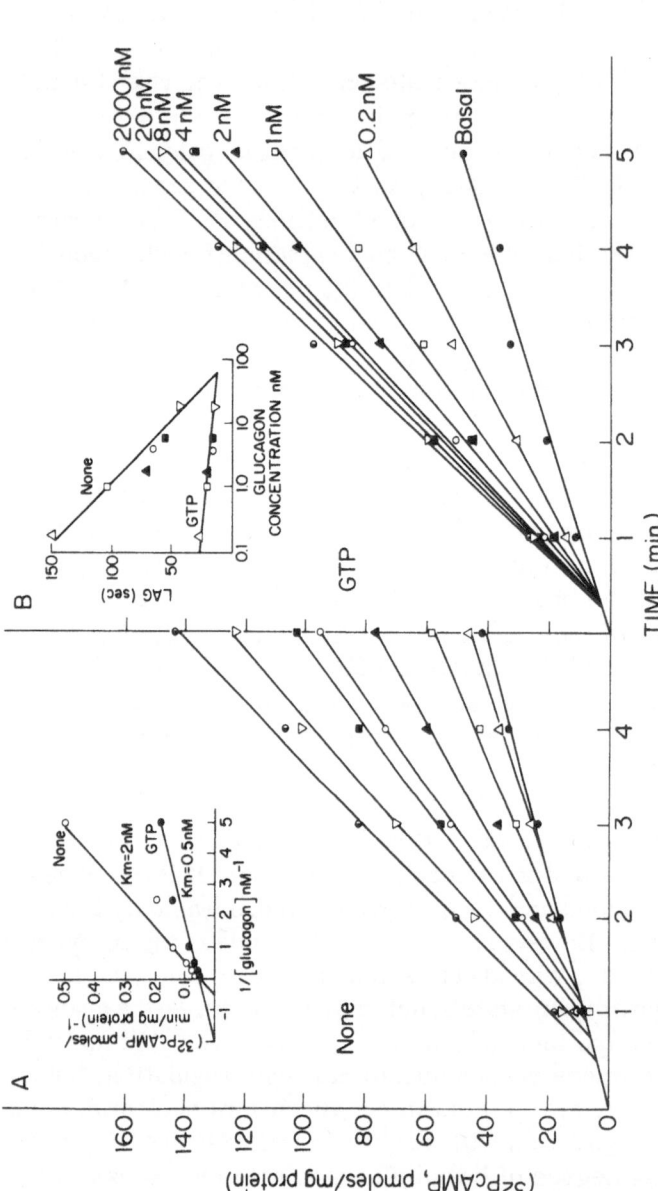

Fig. 71. Time course of glucagon action on adenylate cyclase in the absence and presence of GTP. Adenylate cyclase activity was measured with 0.1 mM (a^{-32})App(NH)p as substrate in the absence (panel A) and in the presence (panel B) of 0.1mM GTP. The concentrations of glucagon are shown in panel B; the same symbols for the respective concentrations are used in panel a. Rates were obtained from the slope of each curve. The increase in rate due to addition of glucagon was used in the Lineweaver-Burk plot (inset A) to determine the apparent K_m for glucagon. The lag time required to reach constant rate at each glucagon concentration was obtained from the intercept of the rate curves with the curve given by basal activity. Lags were plotted as a function of glucagon concentration (inset B).

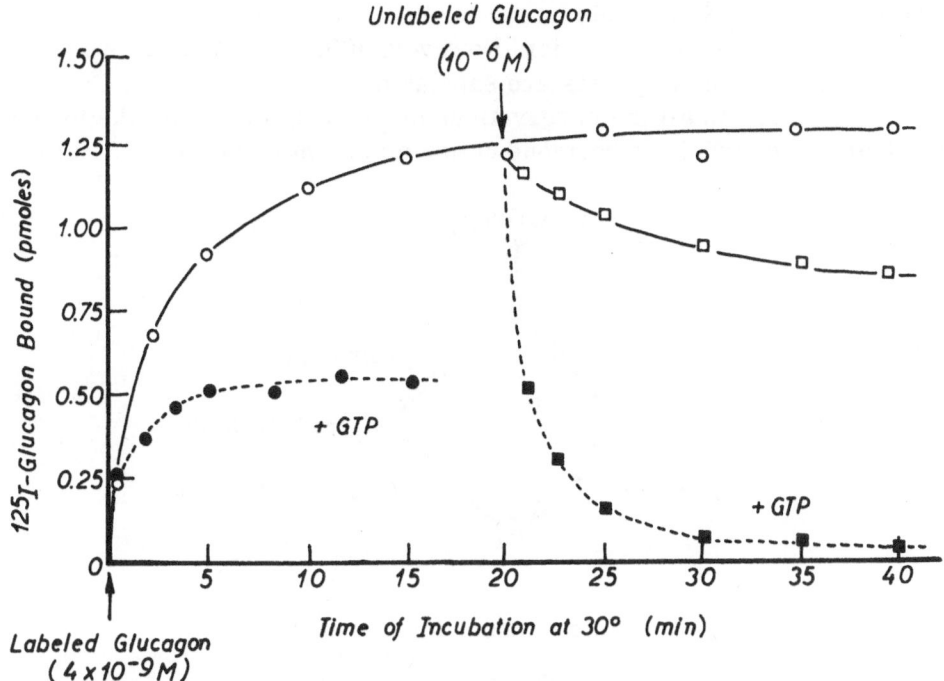

Fig. 72. Binding of 125I-glucagon to hepatic membrane and its dissociation

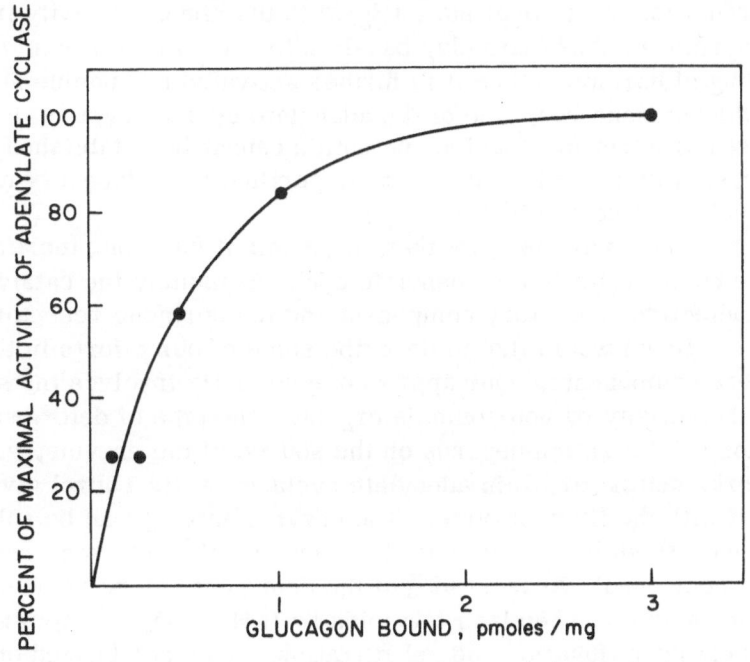

Fig. 73. Adenylate cyclase activity as a function of bound glucagon.

the activation depends not only on the levels of receptor occupancy but also on the nucleotide concentration. However, 100% activity is reached only when all of the binding sites are saturated.

Based on all the observations discussed, we would like to propose a four-state model for hormone action as shown in Figure 74. The

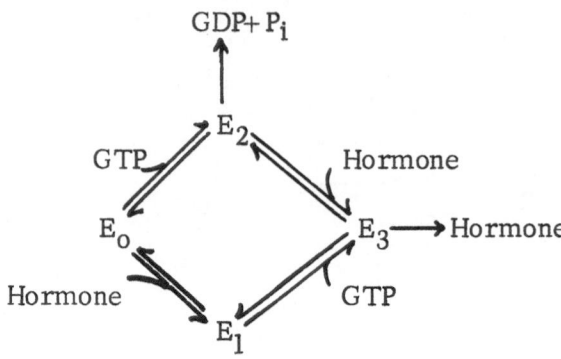

Fig. 74. A four-state model for hormone action.

binding of guanine nucleotide to the basal state (E_0) of the enzyme transforms the enzyme to a more active state (E_2). This state can go through another transition to the final state (E_3) with the highest activity in the presence of hormone. Alternatively, basal state can go to a new state (E_1) upon binding of hormone, then it is further activated by guanine nucleotide. Whether the hormone response of the adenylate cyclase system has an absolute requirement for guanine nucleotide cannot be established until the ATP preparation used as substrate is purified to be free from any guanine nucleotide contamination.

The study suggests that, in terms of function, there are three distinctive components in the adenylate cyclase, namely the catalytic component, nucleotide regulatory component and the hormone receptor.

Now I would like to describe some of our efforts in the isolation of these components. Our approach essentially involves the solubilization of the activity by non-ionic detergent. The type of detergents most suitable for solubilization depends on the source of the enzyme, e.g., digitonin works better for brain adenylate cyclase, while Lubrol gives the best result with the liver enzyme. The enzyme activity can be solubilized with at least 80% yield and the stability shows an absolute requirement for sulfhydryl compound. At least 99% of the detergent can be removed and the enzyme can be purified by a few folds by $(NH_4)_2 SO_4$ precipitation, sucrose gradient centrifugation and gel filtration. This solubilized enzyme is still responsive to NaF and to guanine nucleotide but not to glucagon. The addition of fluoride or nucleotide to the solubilized enzyme not only elevated the enzyme activity but also markedly enhanced the stability of the preparation. When Gpp(NH)p was added during the solubilization and thereafter, the half-life of the enzyme activity at 4^0 was markedly extended. Now we are in the process of purifying this enzyme which has been stabilized by guanine nucleotide, with ion-exchange and affinity chromatographies. Our effort on the receptor isolation has been rather limited up to now. We did observe that after the removal of detergent, the solubilized adenylate cyclase system binds glucagon although with a lower affinity than membrane preparation. However, since the solubilized enzyme does not respond to glucagon, it is difficult to know the relevancy of this hormone binding protein. Therefore, our objective now is to first isolate the catalytic and the nucleotide regulatory components and hopefully it will be possible some day to use this isolated components to assay for the hormone receptor.

KAHAN: What is the final result of this interaction of hormones or "hormone-like" substances binding at the cell surface thereby altering the activity of certain cell surface components and then eventually yielding effects on the interior? Jerry Price, could you comment on surface membrane regulatory factors affecting granulocytopoitis.

PRICE: For several years now, a large group in Toronto has been studying growth and differentiation of hemopoietic cells. Several years ago Till and McCullock reported a pluripotent stem cell capable of giving rise to cells making up peripheral blood. Today I will discuss one of these lines of differentiation, that principally of granulocytes.

In this pathway, there exists a progenitor or committed stem cell for which we have a colony assay. We call this cell CFU-C. This progenitor is dependent upon a factor we refer to as colony stimulating activity, CSA, for growth and differentiation in vitro. (We use peripheral leukocyte conditioned modium as our source of CSA.) We are studying the differentiation of granulocytes from these committed stem cells in human marrow. Human marrow differs from that of mouse and other species in that there exists a population of cells in the marrow which, themselves, produce the colony stimulating activity. Messner (Blood,1973), has reported a method for removing these CSA producing cells from the marrow, leaving the stem cells, that is, the CFU-C free of these cells. The method took advantage of the fact that CSA-producing cells are adherent.

These observations led us to consider a hypothesis proposed by Rubin in Major Problems in Developmental Biology, 1966. He proposed that cells that demonstrated density dependence or were dependent on condition medium may be responding to membrane components or molecules existing in equilibrium between supernatant, culture medium, and the cell membrane of other cells.

Recently, the Pellegrino's, Reisfeld, and Kahan have observed that spent or conditioned medium contains sources of HL-A antigen. This procedure involved centrifugation at 160,000 x G for six hours and resulted in a pellet of membranous debris containing the histocompatability antigen. When we subjected our leukocyte conditioned medium to such treatment, we observed a greater than fifty percent loss of colony stimulating activity. The pellet, however, by itself could not compensate for the loss of activity seen in the supernatant medium. However, on reconstitution, we were able to fully reconstruct the colony stimulating activity by appropriately mixing the membranous debris back with the supernatant fluid. We interpret this to be either a result of the pellet containing particulate material as a reservoir of CSA, or in some way enhances the molecular colony stimulating activity.

We directly tested this hypothesis by isolating membranes from human, normal peripheral leukocytes and fractionating them by a modification of the method of Burnette and Till (J. Membrane Biol., 1971). This procedure leads to three fractions: One which is a cytosol. The other is a pellet containing nuclei endoplasmic reticulum, golgi, mitochondria, and other cellular fragments. The plasma membrane partitions at the interface between polyethylene glycol and dextran. This isolated plasma membrane, after solubilization with SDS, was shown to contain the greatest amount of colony stimulating activity. In four separate experiments, a 10 to 100-fold enrichment in the membrane fraction relative to the other fraction of colony stimulating activity was observed.

We know from purification of leukocyte conditioned medium that four molecular species exist. These range from 70,000 molecular weight to 35,000, to 15,000 and very small molecule, less than 1,300 molecular weight. The higher molecular weight substances are at least protein. We did not yet know whether or not any of them are glycoproteins. The low molecular weight material is probably a hydrophobic peptide. Upon solubilization and examination of membrane fractions, all four molecular species are seen.

The disease known as leukemia represents a possible deviation in normal differentiation. We examined leukocyte conditioned medium prepared from several patients who were untreated for presence of these colony stimulating activities. To date, in ten patients we have only been able to detect one apparent molecular species of the high molecular weight type. Although low molecular weight CSA exists in all of them. However, when simultaneous direct preparations from membranes are done at the time of production of leukocyte conditioned medium, one sees that all three high molecular weight species exist in the membrane, although only one is seen in the leukocyte conditioned medium.

There are several hypothesis that might explain this. One which we put forward today is that these colony stimulating activities in normal leukocytes are exposed to the surrounding medium and can be shed. In patients with leukemia, transformation events may sequester these molecules such that they are now not externally exposed nor available for shedding as leukocyte condition media. This hypothesis is currently being tested in our laboratory.

KAHAN: Can we suggest that there exist within the cell membrane some factors which are stimulatory and can induce all proliferation. Also within the cell surface might be substances whose action would result in an inhibition of proliferation, namely chalones. Fritz, is it possible that the MLC phenomenon might depend upon the local release of a material like this which translates the membrane recognition message into proliferation.

BACH: Yes, certainly.

RANNEY: I will briefly describe an inhibitory material that can be obtained from the conditioned medium in which lymphoid cells have been cultured or incubated. In our system, splenic lymphocytes from mice, rats and several other species, are cultured for various time intervals. The supernatants are removed and added at serial dilutions to isologous lymphoid cell targets. These targets can be either B or T cells, but the effect is predominantly on lymphoid cells as opposed to fibroblasts or other nonlymphoid cells. Addition of the unfractionated supernatants to lymphoid targets produces a marked inhibition of the incorporation of tritiated thymidine, and also decreases the mitotic index and the nuclear labeling index by autoradiography. The UM-2 Amicon filtrates (consisting of species below approximately 1000 in molecular weight) are more inhibitory than the whole supernatants. On sephadex G-10, we obtain an approximate figure of 600 molecular weight for the peak inhibitory activity.

The interesting point relevant to your presentation is that the material retained above the Amicon filter produces considerable stimulation at all dilutions. This segregation of inhibitory and stimulatory activities by molecular weight could possibly account for the differences you see between the high and the lower molecular weight species of CSA.

KAHAN: Jerry is shaking his head. Can you give us a comment on that?

PRICE: When we test any of these molecular weight species for thymidine incorporating activity on leukocytes, leukemic cells, on L-cells, or

on Chinese Hamster Ovary cells at concentrations from anything equivalent to six-tenths percent of what was present in the original supernatant
through 600 percent equivalent, we see no thymidine incorporating activity.
That we know for these factors. However, we do know that we can -- and
this is very preliminary in the purification procedure we have for the CSA
-- separate out a molecule of high molecular weight, in excess probably
of 35,000 that will stimulate thymidine incorporation. With regard to low
molecular weight species, it came quite rapidly and it is a very different
topic. I am not so sure that it qualifies as a chalone, there is no specificity which is normally associated with chalones. I didn't understand
whether or not you said there was any characterization of that small
material. The molecular weight is not that far out, and thymidine has
sometimes been mistaken for a chalone.

RANNEY: We have done some initial characterization. The inhibitory
activity has been purified to the extent that only about 1/10 as much of the
factor (by weight) is required to produce a 50% inhibition (ID_{50}) as is
required for known concentrations of cold thymidine. Therefore, even
though the factor has been only partially purified, it is a reasonable assumption that it is not thymidine.

PRICE: What is it?

RANNEY: I can't tell you at this point what it is. However, one more
comment I would like to make about the system is that the inhibitory activity is obtained predominantly from the supernatants of higher density, unstimulated cultures. If the spleen cells are stimulated either with mitogens or allogeneic cells, the release of inhibitory activity is almost completely suppressed.

PRICE: In contrast for the CSA stimulation, PHA boosts considerably the
amount of activity released, and that is true for all fractions of the high
molecular weight CSA's.

RANNEY: That is compatible with the effects we observe following stimulation, but it does suggest, as you observed before, that the decreased
stimulatory activity present in your lower molecular weight fraction is not
due to the presence of chalone-like factors.

KAHAN: The PHA leads to release of the material which perpetuates the
PHA effect. In any case, in view of the hour, we had best adjourn. The
range of biologic activities of surface membrane components which were
discussed this afternoon certainly demonstrated that each system has
unique methodology which may be useful in unreavelling the mysteries of

the biological function of the material. Hopefully, the cross-fertilization of this multi-disciplinary apposition of approaches may afford new insights into the problems each of us face in our given area of cell surface research.

SESSION V

PERSPECTIVE OF CELL SURFACE
STRUCTURE AND FUNCTION

Heterogeneity of membrane phase - Macromolecular interactions -
Profile o f surface membrane complex - Hierarchy of membrane
forces - Receptor mobility - Membrane states during c e ll cycle-
Assembly and fu n c tio n of hi sto co m pa tib ili ty a n ti ge n s -
Polymorphism - Molecular degeneracy.

SESSION V

A PERSPECTIVE OF CELL SURFACE STRUCTURE AND FUNCTION

KAHAN: I think it is appropriate to have Dr. Edelman bring together some of the information that we have learned here, into a general construction, with his unique perspective.

EDELMAN: Summarizing a complex subject and a complex meeting is a bit like playing a violin, and like violinists, summarizers come in three classes: Those who can't do it at all, those who do it very badly, and those who do it very well. I hope I will be able to advance to Class 2 without too much pain to this group.

The text I shall follow is that of Einstein, who said that is the theory that determines what you can observe. Therefore, what I am going to do, in terms of the questions raised by this subject, is to theorize in a way that would ordinarily be considered scandalous. In fact, there are so many incomplete aspects of the subject, that if one does not do this, discussions of the subject will degenerate into a mere catalog.

Still, it is the facts themselves that are astonishing, and therefore, where we do have facts, I shall try to emphasize those that I think are particularly significant.

The key problem we have been facing, I think, is that the cell surface-membrane complex is very complex indeed. It consists of very many different kinds of molecules distributed in heterogenous phases, and we have very little information on their direct linkage. First, I shall attempt to summarize some of the boundary conditions that relate to these parts. Then I shall discuss a minimal model of the membrane-surface complex in terms of some major cellular functions. Finally, if time permits, I will touch upon the problem of cell-cell recognition, a field that might be called, with all due respect to Moscona, para-immunology.

This field does not yet have the discriminatory situation that once existed in immunology. That was a situation that Niels Jerne described as "cis" and "trans." But we can expect to have differences in points of view and background and, in fact, that is one of our difficulties in talking to each other. Nevertheless, I think one of the most cheerful aspects of this meeting and similar meetings that I have had the privilege of attending is that there has not been this division, this sharp division, as to types of background.

There is a story about two brilliant British boys who grew up in school together, hated each other roundly and competed with each other all through school. One of them became an admiral and one became a bishop. One day they encountered each other, quite by accident, in Waterloo Station. The bishop eyed the admiral and the admiral coldly stared back. Finally, the bishop walked up to the admiral and said, "Pardon me, conductor, is this the way to Picadilly."

The admiral looked him over and finally said, "Yes, madam, but in your condition, should you go?"

I don't think anything of that sort exists here, although I believe there is a great difficulty in reconciling differences in methodology. I hope that we can iron out some of that today.

It does not take very much insight to see that the evidence suggests that there are in the surface-membrane complex interactions of at least three components connected in a variety of ways. There are lipid-lipid interactions, lipid-protein interactions, protein-protein interactions, protein-carbohydrate interactions, carbohydrate-carbohydrate interactions, and finally carbohydrate-lipid interactions.

Some of these occur via weak forces and some are covalent interactions. All of them are distributed in a complex of phases. In view of this situation, I shall limit in my discussion to those receptors that penetrate the membrane. In other words, forgetting all of the details for the moment, I am going to consider any glycoprotein receptor that penetrates the membrane and ask in fact what kind of forces it might be exposed to. This means that the forces shall be evaluated in terms of the various interactions, but in a very naive way.

If we consider this simple diagram (Figure 75),

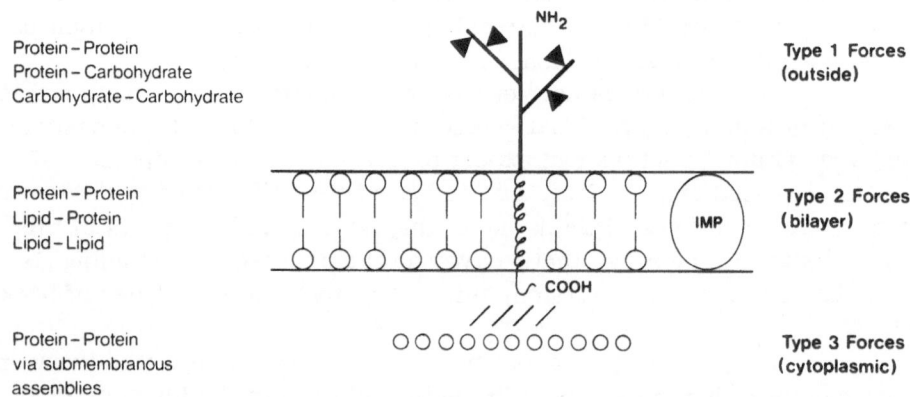

Fig. 75. Forces confronting a glycoprotein receptor molecule which penetrates the membrane.

it is fairly clear that there must be some kind of hierarchy of forces, and a variety of different interactions in each of these various phases.

I don't think anyone has a precise estimate of the order of these forces relative to each other. There are some interactions of a lateral type over here at the surface (type I) and there are certainly interactions of the receptors meeting within the lipid bilayer (type II) and also probably interactions in sub-membranous assemblies and structures (type III).

All of these interact with each other in a complex way, but I would make the guess that forces of type III are going to dominate most of the significant distributions of receptors and their relation to the biological activities of the cell.

Nonetheless, I must add that forces of type II obviously can play a strong exclusive role. That is to say, they may determine the conditions under which a series of forces cannot push a receptor into a certain region, for example, a solid "island" of lipid.

Before I continue, let me try to take this very complex situation involving phase heterogeneity and very complex interacting molecules and mention what I think have been some of the key simplifications in the field. Even though we don't have a quantitative estimate of these interactions, I think that these simplifications do give us a framework within which to proceed.

I think one of the key simplifications was laid down with McConnell's observations on the fluidity of the membrane lipids. It seems curious to me that this was not simply deduced from the biological behavior of the cell, and cell motion. But, in fact, it took a very complex apparatus and a profound mind to see, in fact, how important this would be. This key observation was complemented by those of Frye and Edidin who took a different approach and demonstrated, I think persuasively, that there could be movement of cellular antigens within the plane of the membrane.

Both of these observations provided what I shall call the ground for a plan of the surface - membrane complex. But I think it would be misrepresenting the field if people thought that the subject just stopped there. I think that there are a lot of people who think the subject just stops there; that the surface is simply a kind of layer of melted lard in which corks, representing receptors, are bobbing up and down or moving about.

There are, in fact, two other developments in the dynamics of the surface membrane complex that seem to me to be of great interest. Both are based on the notion of fluidity and receptor mobility, but they point to the existence of higher order modulation events. These developments include the findings by Raff, Taylor and his associates and Karnovsky and his associates on the capping and patching phenomenon, and the work of Yahara and myself on the restriction of mobility by certain lectins. Both of these observations, I think, point to higher order structures that are interacting with membrane components, and I think they will be of great significance in interpreting the functional aspects of membrane - surface interactions. In particular, they point to the possible existence of dynamic submembranous assemblies that are linked to the surface components.

Thanks to all of these ideas, we have some reasonably simple models of how the membrane might work in terms of its surface molecules and submembranous interactions. Completing this picture is what I would call the structural basis, with the red blood cell as the key example. It seems to me the work of Steck and Marchesi is particularly significant. The isolation of membrane proteins and related molecules, and the identification of glycophorin and spectrin as discrete chemical entities, provide a degree of concrete knowledge that, unfortunately, is not possessed for any other cell type so far. These studies provide us with an example, both intellectually and practically, that we must in fact follow up and admire. They not only provided a paradigm for the fractionation of membrane components, but also allow us to talk more concretely about the classes of protein -- protein interaction.

I think all of this now allows us to take a little bit of liberty. We can say that with the red cell proteins we don't understand the function but know something about the structure and conversely with lymphocyte and fibroblast proteins we know something about the function but not much about the structure. Nevertheless, the combined information puts us in a position to say that we know something about the plan of the surface membrane complex.

In fact, it is not a static plan. As Marty Raff and Morris Karnovsky have pointed out, it is not clear whether the plan involves random, quasi-random, or perhaps locally ordered interaction, as one observes the surface from the outside. This is a question that remains to be answered. The fact that capping can be used as a tool has suggested that there is a reasonable amount of independent motion of different receptors under certain constraints, and I believe this is a very important point that should be made for those who doubt that capping has certain physiological significance; it certainly has laboratory significance.

This is all very well, but what I think we really do not have is what I will call the profile of the surface - membrane complex. It is schematically shown by this slightly absurd figure (Figure 76).

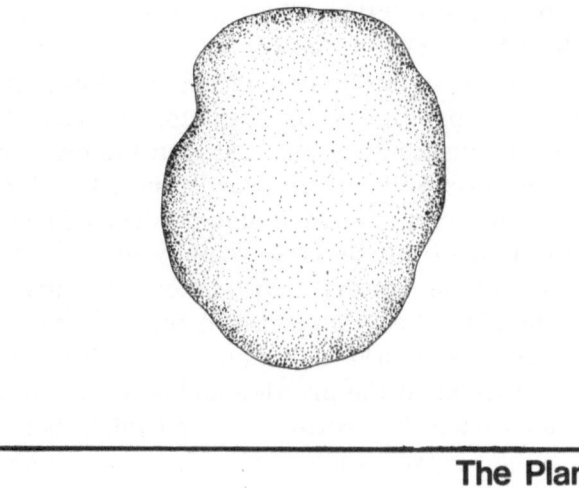

The Plan

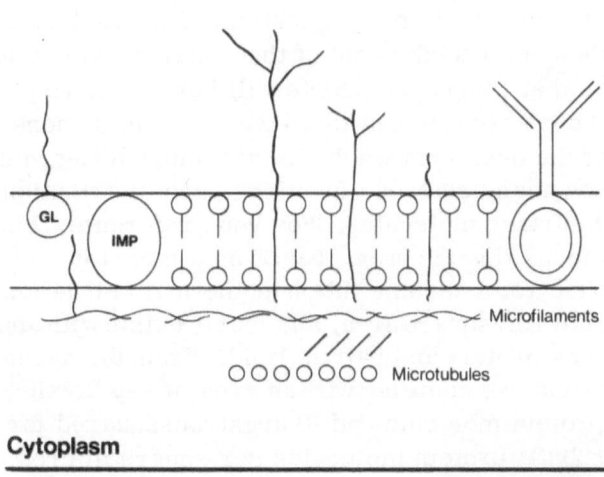

The Profile

Fig. 76. Postulated profile of the surface membrane complex.

H e r e I think that the major contribution, the only one that gives us any
k i n d of knowledge has been made by Dan Branton in h i s studies of i n t r a-
membranous particles. We do know something about this aspect of the profile; sometimes by deduction, sometimes by inspection.   This is the first
clue, I think, in terms of the relationship of the structure to the chemical
material that Marchesi has isolated. But I think, in fact, that given the com-
plexities we h a v e been talking about a n d the kind of interactions, we are
miserably ignorant of t h e profile.   For example, w e don't know anything
about the level of the various kinds of molecules in the membrane. That is
to say, what constraints are on them in terms of the kind of molecular sub-
structures that m i g h t interact with each other.   The constrains o n these
structures include their flexibility and the capacity of various components of
the structures to bend down and interact with other structures.   We don't
e v e n know what the heights of all these structures are, a n d how far they
extend from the surface.   So I t h i n k that among the things we have to develop are new ways of looking at the profile, and new ways of d e d u c i n g
some of the interactions from the profile.   I don't think that s i m p l e electron microscopy is going to give us much of a clue, unless someone thinks
of something very, very bright.   At the levels of resolution and discrimi-
nation required, it won't be possible to use simple staining m e t h o d s to
bring out a picture of these levels of interaction.

        W i t h this background, we can perhaps focus the discussion on
some questions. If you ask too many question, of course, everything disin-
tegrates.   Nonetheless, I think some of them may provide a focus f o r the
general discussion that I hope you people will have. I have placed some of
the questions that I think a r e relevant to all these considerations on the board.

        One of the questions we don't know much about simply in terms
of taxonomy, is how many classes, functions, and distributions of classes
there are for cell surface molecules. For your reference, I have put down
some numbers to focus this question, based on a monster cell suspiciously
like a small lymphocyte. I assume this hypothetical cell is six microns i n
diameter, has 100 micron square area, and that it exists with approximately
half of its surface as protein and half as lipid. From the assumptions about
the diameter and area, we come out with an area of say 2.7×10^3 angstrom
squared for each protein molecule and 70 angstrom squared for the lipid. The
surface has about 2×10^4 protein molecules per square micron, a n d 7×10^5
lipid molecules per square micron. Since this cell looks suspiciously like a
lymphocyte, I assume there are 10^4 intramembranous particles (or IMPS)
per cell.   T h i s is grossly off, because I have little knowledge, but I have
assumed there are about 5×10^6 (possibly 10^7) carbohydrate chains associated
with the cell surface. Just to show you the great range of numbers one could
have, I've assumed that the number of tetrodotoxin sensitive sites in a neu-
ron is something like 13 per micron square, and potassium sites, something
like 800 per micron square.

As one focuses on that first question in terms of these numbers, we see that we are pitifully lacking a specific taxonomy of molecules for a given cell type. We do have certain estimates, more or less, of the number of specific molecules per cell. For example, if this is taken not too seriously, 5×10^5 H-2 molecules, perhaps 4×10 TL antigens, 4×10^5 Thy 1 and 2, 2 times 10^4 immunoglobulins and for Con A, a grab bag, 2 times 10^6. We remind ourselves that this cell probably has the order of 5 times 10^6 protein molecules all together.

Now the question I am asking is, if we could name them all, and if we were so lucky as to be able to relate them to functions, what would the class distribution look? Would it look like this: a major class, performing some structural function, and then a large series of little classes with different activities. Or would there be bimodal or trimodal distributions with a great deal of discrepancy in the numbers of types represented? Essentially, the number of different classes of receptors and the frequency distribution in each class as well as the relationship of the classes poses an evolutionary problem for which we don't have enough information.

Second, I think we have to ask, how many of the different receptors, in fact, go through and through the membrane. Everybody draws the receptors differently and people have different prejudices about which ones are closer to the inner lamella or outer lamella and which go zooming through the entire bilayer. How many of these different kinds do we have, in fact?

I think I have already touched on the third question which is, how many actual levels are there in profile? Are there proteins that stick out with their carbohydrates extending very far and others with their prosthetic groups clustering near the membrane. What provides the vertical stability for this kind of situation, if it indeed exists. Is it only the protein-lipid interaction?

Fourth, I think in view of the very interesting remarks of Pedro Cuatrecasas and Dr. Hakamori, we have to consider another kind of structure very much more seriously: that of the glycolipids in the membrane In particular, there is the possibility that their position can be altered by certain mechanisms of the kind that Pedro has mentioned, namely inversions of a curiously specific type. It is of general interest whether, in fact, their positions could be altered by convection or by countercurrent movements in the lipid bilayer.

Fifth, I think that the intermembranous particles have to be discussed in terms of questions of specificity. Are the intramembranous particles architectural elements only? Are they, in fact, there because it is required when two membranes fuse, for example, that a certain structural stability be imposed at that temperature on the phase-separated complex of islands and seas? Alternatively, as suggested by Dan Branton, do they carry specificity, that is, do they represent molecules that stick out and have specific functions in the outer world?

Sixth, I think a similar question has to be asked about gap junctions. I think that here we are particularly fortunate because it has been shown that we can actually isolate the molecules of which they are composed. They are without a doubt highly specialized surface components. Are they merely architectural (which is not to diminish their role), or do they, in fact, show specific interactions from cell to cell? Perhaps we can get back to how we can answer questions of this type at the very end of the discussion.

Now, the seventh question, that concerning the anchorage of receptors, has come up repeatedly. All of us sort of throw up our hands when it comes to elaborating just exactly how this is to take place. I have just added to the confusion by suggesting two more categories, that I think have seriously to be considered. The first I will call the piggy-back anchorage, the notion that, for example, the Fc receptor of a lymphocyte is piggy-backing the CH3 domain of an Ig molecule. The second is curious: no one has actually ruled out that there could not be covalent linkage of some of these molecules to whatever structures are in fact embedded in the membrane. Question eight is one that has been discussed speculatively, but I believe we have no information on it whatever. This concerns the exact nature of the linkage to submembranous assembly. As you have heard, these submembranous assemblies are very complicated indeed. Each of the suspected actors in the play of submembranous structures is being studied by a group with perhaps different motives than the present group. One can take some hope that the studies, for example, of those people interested in microtubules and of those interested in microfilaments will converge upon ours.

Then I think it is very important to consider the question of modulation, number nine on the list. We have to consider it hypothetically and actually in terms of the kind of deductions that have gone on in this meeting. We have to try to make a hierarchy of forces, to make a guess about the kinds of forces involved if we assume that such an assembly is arranged more or less like that in Figure 75, and we must discuss whether phase modulation can dominate in one situation, and whether macromolecular assemblies can dominate in another. Above all, we must see whether modulation is a selective or general event. By this I mean whether, when you have modulation, it is all or none, or whether certain selected receptors can be modulated while others are free to move. I think this will be very important in interpreting the functions of the cell.

Number ten seems to me to be a key question. What molecules of the membrane are truly specific? Let me define what I mean. I don't mean, for example, that the glycolipid that binds the cholera toxin is not specific. It is specific in a chemical sense, but I mean specificity in the sense that we associate the term with enzymes and antibodies, namely, a range of clearly determined specificities carrying out a variety of functions and potential functions. Here we have to discuss whether, like other systems, proteins are not the best candidates, or whether some of the other kinds of molecules couldn't in fact take on a range of specificities. Personally, I would bet on the proteins by a wide margin.

Finally, I think, we have to consider this whole matter of cell receptors in terms of their synthesis. We have to ask particularly about the control of synthesis, about which I think you can make two extreme models. If you have a Raffian situation (or a Raffish situation) capping away one receptor, and then resynthesize that particular receptor, what in fact is the cell doing? Is there a specific feedback of information that says: make more of that receptor? Or are you, in fact, in a steady state, and have failed to observe that when you do this, you stimulate the synthesis of a great variety of receptors, with non-specific exclusion of all of those other than the capped receptor? I don't believe anybody has discussed this in any great detail.

Let me turn now to a matter that has to be considered in a larger sense, one that is so large, of course, that in a meeting like this, we don't really discuss it in detail: the relation of all of this to several fundamental cell biological functions. These functions include, for example, cell division, cell movement (which, by the way, is a very big grab-bag, since the movements are really whole classes of different kinds of movement, differently generated), cell-cell recognition, and cell death. This last subject is too depressing, but it has been discussed by Gordon Tomkins in a provocative way and is very important. At this particular meeting, we have enough on our hands, so we will just not consider it.

What is the relationship of some of the surface models to some of these functions? There are several things on this subject that have been kicked around in the literature, I think, and oldtimers like Aaron Moscona can comment about them. He, I think, will agree that people who have had his kind of experience, know that cells have a curious kind of interaction with surfaces that just refuses obstinately to be explained in terms of known interactions or structures.

If you assume a general kind of interaction, that is an equilibrium, in which the cell has a bunch of receptors that have been brought together in a metastable configuration sufficiently close to be considered multivalent, then the free energy of bending can be represented by $-\Delta F = RT \Sigma \ln K_i$ for different receptors, where K_i is the association constant and $-\Delta F = RT \ln K^n$ for identical receptors.

This means, that if a cell is on a surface and the receptors are in a cluster, then the likelihood that they will all be free and the cell be released may be very small indeed.

Although this is significant for known specific receptors, I personally don't believe that this can explain the stickiness of the cells to

various surfaces because I believe there must be a lower order of interaction that we know nothing about. I am going to opt for the carbohydrates
as having something to do with this interaction, and I am not going to call
it non-specific, because I believe that only clouds the issue and just means
we don't know what kind of interaction it is.

Nonetheless, those of us who have tried to fractionate cells
have run up against this problem in a most disgusting way. You hook a cell
specifically onto a matrix that you have designed to interact with its known
receptor. After a suitable thermodynamic calculation, you add free ligand
to break this specific interaction, and you find the cell won't let go. We have
tried to fractionate red cells on affinity columns, using Con A. The red
cells will stick specifically as proven by the fact that the cells do not stick
in the presence of the competitive inhibitor α-methyl mannoside. But once
the cells were stuck, you could pour in all the competitive inhibitor you
wished and they would stay there, until you did one thing. If you lowered
the osmolarity to almost the critical point, and then added the inhibitor, the
cells would come off, but not if you had either condition alone. I don't know
what this means but suspect that once a cell binds by one set of specific receptors, there are probably additional interactions with the surface via
other molecules.

Let me now say something about modulation in this connec-
tion because I think it is very important in terms of cell-cell recognition and
cell movement. Is it possible that modulation can be selective? To my
knowledge, there is only one set of observations that indicate that it might
be so, those of Berlin and his associates. In Berlin's experiment, transport
receptors stay put on the surface of polymorphs while other receptors such
as those for Con A were phagocytized except when colchicine was added, in
which case both receptors went in.

What Raff and Yahara, and people interested in the problem
are looking at, may be a phenomenon which is generated by their experimental protocol. Modulation may affect all receptors without any selectivity in
restricting the mobility of some while letting others move. If that is so,
there is nothing that you can do about it, and the subject stops with a dull
thud. But if, in fact, their conditions simply cannot tune in on the various
states of the modulating assemblies, then they are failing to see that the cell
has an apparatus which can hold some receptors still and allow others to
move or to be moved. I think that if such a mechanism exists, it must be
pretty important when you consider the multicellular attachment problem
even independently of the problem of the gap junction.

I am now going to provoke comment by saying that I think that the problem of cell receptor mobility is inseparable from the problem of cell movement itself. It behooves us to consider that perhaps the first things to make contact with the outside world are some of the surface glycoproteins, particularly the glyco part. Although I personally do not believe sugar-sugar interactions have a large range of specificities and believe dogmatically with no real evidence, that only proteins can have the kind of specificity we would like, nonetheless, perhaps the first thing that is seen in the outside world is a carbohydrate which can, of course, interact with another carbohydrate on another cell in attractive or a repulsive way. This same carbohydrate "foot" may be the first thing to interact weakly with hydrophilic surfaces.

It may be such a glycoprotein sticking to the modulating assembly, which is the key assembly to focus on in analyzing cell movement not to the membrane, as Abercrombie an others have done in discussing cell movement. They have looked at conservation conditions for cell movement and translocation, and have made use of tracer particles. From their observations, it seems one might have to synthesize membrane at an inordinate rate. I believe such models have to be reformulated in terms of what we have been talking about here, namely, that the receptors play a much more important role in the generalized and global motions of the cell than has been realized. If this is so, and if the interactions are highly cooperative as our experiments on cells attached to fiber suggest, then attachment of merely one small region of the cell can alter the entire behavior of other regions. This cooperativity could be negative or positive and would certainly have to be considered in discussing the factors determining the types and directions of cell movements.

In this connection, I think it is worth pointing out that there must be a whole series of equilibria to consider. The membrane may be there mainly for separating inside and outside and the linkage of the through and through receptors may be the key determinants of these quilibria. There is an equilibrium between the receptor and an outside ligand, that between the receptor and microfilaments or microtubules, and finally between these structures and intracellular small molecules. With such equilibria, there would be no difficulty in principle in constructing a cooperative network although, of course, it would be very complex. Perturbation of all of the equilibria could come from inside or outside of the cell (see Figure 75).

Finally, in terms of what I believe to be an attractive model related to the question of the stimulus for cell division, you might even have the microtubule as a buffer of biochemical reactions involving messengers of a kind that go to the nucleus that Tomkins spoke about. This would allow for mechanochemical transduction of signals of a variety of types.

In relation to this next subject, which is the subject of growth control that has come up here in various guises, this is a much more attractive model than a model that simply involves transport of molecules through the c e l l membrane. It is my personal belief that transport must play an enormous role, but a simple model which says that you dump a messenger outside or push it inside does not have the kind of control that would be required. The phenomenological observations of biologists and cell biologists, that there is s o m e relation between cell movement, cell growth, a n d the speed at which certain control events seem to take place would seem to require a structurally specific assembly in addition to transport.

I don't k n o w if it has been confirmed, but I b e l i e v e t h a t Piatagorsky showed that lens cuboidal epithelial cells could be put in culture tissue and their shape altered. He added insulin to cells blocked for their protein synthesis and observed i n short order that the cells elongated and that there were long microtubules, in the direction of elongation, assembled under the cell membrane.

I believe to control s u c h an event, it is a much more reasonable thing to have an ordered assembly modulating transport events rather than simply to have a pore which just leaks some stuff out. This is a question that remains open for discussion in relation to the control of cell division.

The c e l l division problem has been touched upon by Tomkins, Cunningham and others here, a n d I believe what is interesting about that problem at the present moment (to be analyzed in relation to our knowledge of various cells) is the question of decoupling of the stimulation from specificity. People have gone very far in proposing, for example, that immune stimulation has nothing to do with the antibody receptor, but only with mitogenic properties of the antigen and that all antibodies do is concentrate mitogen at the cell surface. I think that idea, which I reject, in fact, has to be distinguished from the more agreeable idea, that there is no need for the cell to couple the specificity of the receptor per se to the mechanism of stimulation. That is to say, there must in fact, taking an immune example, be some final common pathway, possibly inside the cell, or under the membrane which is mediating all the alphabetically different ways in which we tickle the cell. This pathway will be generally shared by antigens, lectins, hormones, and so forth. I think Tomkins, w h o unfortunately had to leave, generally agrees with this idea.

The second thing that I think is interesting about cell division was mentioned by Cunningham. There is a new attitude toward the cell cycle which is that it is not a continuous fixed cycle but is rather in a series of states like the A state and B state of Smith and Martin.

It is intriguing that Wang, Gunther and Cunningham's experiments, and those of Temin, strengthen the possibility that, instead of defining one cycle, there are some probabilistically distributed set of states in what is called G-1, and a defined set of states related to S, G-2 and M. I think if this is true, one is obliged to ask what evolutionary function and fundamental physiological functions does it serve to have resting cells each with a different G_1-S gap.

I think this, in fact, is a question that first has to be asked in relation to the cell surface. I believe it was Cunningham who said yesterday that there were experiments to suggest that serum was not required and some to suggest that cell contact was not required, but I didn't hear the statement that diffusion of other products from second cells in addition to the ones that you were adding was not required.

I have assiduously avoided the whole question, you will notice, of second signals, the kind of things that immunologists have talked about. I believe that, although there is much to be said about this in terms of cell-cell interactions, of co-factors and co-mitogens, the problem is complicated enough without bringing in that phenomenon until we sort out some of the events on the cell surface.

Let me turn finally to what I consider one of the most fascinating areas of all, the problem of cell-cell recognition, and the diversity and specificity of cell surface receptors. I am going to make one of those scandalous theoretical excursions now at a rather general level and see how it applies. I believe it is important to relate what we have been hearing about to the chemical facts. We have heard from Aaron Moscona, and all of the people working on histocompatibility and transplantation antigens a great number of things that bear on this question of cell-cell recognition.

I don't really know the field well enough to know whether the conventional model of developmental biologists is one of strict temporal gene programming for single surface products, that are capable, of recognizing cell surfaces in a coordinated way during development. But anyone who has confronted the near-miracle of retinotectal and tectal-occipitocortical interactions which seems to imply the inheritance of a kind of Euclidian geometry, which then can be modulated by interactions with the world during a critical period, is forced to ask himself whether such a naive idea of strict programming is sufficient. So I will begin here -- which may seem far from the problem of histocompatibility -- because I believe that if you don't confront this problem in discussing the relationship of genetics to function, you are really avoiding the issue.

What is the issue? The issue is whether identical twins developing in phase construct a complex organ from dividing and differentiating cells in an identical fashion at the level of their surface molecules.

Let me state that in another way. You could say that cell movement, cell division, and cell death, all appear to be going on more or less simultaneously in some kind of horrendous population dynamics that no one to my knowledge has ever had the nerve to analyze. You might say that the information for all of these processes could be directly programmed in the genetic information; it is at least conceivable that you could do this. But you would have to use a lot of information to do it; you would have to build in an awful lot of information just for safety because any early catastrophe could wipe out the whole system

For the purpose of my scandalous theory, I reject the notion that any such thing is true. I would take the position, in fact, there is no way in which you could develop so complex a set of organs in such a preprogrammed way in view of what we know about the variability, and the number of cell types in the hierarchy. Undoubtedly, you have gene programming but the main point I want to make is that the system of development must be selective and it must be degenerate.

Let us take some hypothetical cell surface receptor and not distinguish for the moment whether there is an external molecule B recognizing in an ABA configuration or there is an A-A' complementary type of recognition. At any time during the cell division cycle and the production of a population which is moving and reassembling according to very complex rules about which we know very little, let us assume that there is more than one way in which you can accomplish the recognition of one cell by another. This requires that there is always a set of molecules which could substitute for some other set, and give you about the equivalent degree of specificity and recognition. This is what I mean by calling the system degenerate; this, by the way, does not exclude the notion of gene programming and control. It just means that in the hierarchy that such control is a coarse control and can have no information a priori about which cell is approaching another cell at any time.

You will see that this replaces the notion of strict programming with a programmed repertoire of surface molecules that is degenerate, and can behave selectively with appropriate feedback control. I don't, for a moment think that there is strong evidence for such an idea, but I should add, that for the moment, I don't think it can be rejected out of hand. Moreover, there are some experiments that can be designed to test it.

So let me finally say something about histocompatibility antigens as one possible candidate for such a repertoire and as an example of molecules on the cell surface, the protein part of which is going to mediate the specificity. I think that Fritz Bach and others of you have given us ample examples of the kind of specificity that can be demonstrated in this system and related systems. It is worthwhile to comment first on how many different histocompatibility loci there seem to be. Fritz Bach, in his review, mentioned that there were at least 30 minor histocompatible loci. That is a large number and the question arises immediately; are they too reflecting surface molecules or are they just other intracellular molecules. If we assume that they are surface molecules, they certainly relate to the first question we asked: How many different families of proteins are there on the cell surface?

The second thing I want to say about this subject is that it is a striking observation, that for every cell, the presence of β_2 microglobulin means that there is an immunoglobulin-like domain on the surface. Whatever one's position about the role of β_2 microglobulin in the HL-A molecule, this is a startling fact and you have really to branch now -- you have two choices. You can either say that β_2 evolved somehow from a gene first developed for lymphocytes, and somehow then got expressed in other cells, or you can take what I think is a more reasonable position, and say that β_2 and immunoglobulins both evolved from a common precursor that was expressed on all cell systems regardless of their state of differentiation. In this connection, it should not be ignored that there are more β_2 molecules on the surface than can be accounted for by histocompatibility molecules, in some cases, at least. This difference must be accounted for particularly in terms of anchorage and also the problem we are now considering.

Let us not bother about that now but take this line of thinking all the way to the end and say that the evolutionary precursor for cell surface proteins included the H2 system as one of its descendents. Let us also suppose that the immunoglobulin system simply represents one of the latest refinements of this evolutionary development. The implication is that there is an evolutionary link between the H2 complex and Ig genes. Gally and I have proposed that the Ig gene clusters arose by gene duplication and chromosomal translocation of the early descendents of the H2 gene complex including the IR genes.

Of course, the key question becomes, what is the function of the histocompatibility system? I'm not going to review the history of the various theories, but I think they can be divided into two kinds. First, there are theories which explain the polymorphism of H2 gene products on the basis of transfer from organism to organism, because that is a simple way to explain how polymorphism is either introduced or maintained or both. They include, for example, ideas that fertilization events rely upon such gene products and there could, therefore, be distortion and gametic selection by these means. There is also the idea that viral transmission of cell surface markers from another individual may be involved.

But I would like to focus the issue on whether the problem of polymorphism can be reconciled by assuming functions within the organism, particularly those related to cell-cell recognition. This is not a new idea and many people have expressed it. Fritz Bach has expressed it, and Walter Bodmer has expressed it, but in different terms. I think it worthwhile in this discussion to consider in detail a particular form which I think would be consistent with selective-degenerate recognition systems. Unlike Bodmer's hypothesis, I would not say that the H2 system a n d the whole complex is linked to something which is a recognizer, but rather that it is a recognizer. Despite the differences in the serologic or cellular tests, I would suggest that all of the genes in the entire complex serve these functions. Moreover, similar functions may be served by gene loci not linked to the H2 complex, such as those of the minor histocompatibility loci.

If we take that p o s i t i o n, then, it seems to me that all w e have to do to give these systems a role in cell-cell recognition that allows for maintenance of polymorphism, is to assume that there is one higher degree o f variability within the gene products of these systems than c a n be detected by your methods.

That does not necessarily mean that you have to have V genes as in the Ig family. It simply means that, in sibling mice that are inbred, despite the great variety of your tests, it is not necessarily true that gene products will be absolutely identical. This is certainly true within the Ig system, despite the presence of all types. It, therefore, seems to me the concept of a hierarchy of variability comes in here. The question is how far down can you carry it and when, in fact, do somatic mechanisms of expression take over that you do not see in your tests.

If we assume that t h e s e histocompatibility products h a v e a level of variability under such a hierarchy, they can all be classified very nicely. Polymorphism would, I believe, be generated in a kind of reverse way, according to class delineations and selections for structures and l i n k-ages of the kinds that Fritz Bach has described.

W h a t I am proposing is the notion that cell-cell recognition may involve molecules of this kind, that it may be selective and degenerate, and the order of diversity may be one higher than the kind you have so far classified. That is easy to say. There are no facts that rules it in or out, but I would suggest that there is an experiment that can be done to help.

For example, what F r i t z could do is take his MLR, or h i s various other cell systems, and test highly inbred sibling mice relentlessly on a mouse-by-mouse basis for cross reactions. I will predict that rarely he will, in fact, see an MLR from identical mice.

I would also predict that different individual embryonic c e l l s from sibs would show diversity of expression during development and that there will be "families" of such cells.

I believe that is a more satisfactory expression of the problem than the very ingenious one that Bodmer advanced: that the polymorphism is maintained by linkage disequilibrium because H2 genes are linked to certain unspecified recognizer genes.

Let me come to the question M a r t y r a i s e d about Aaron Moscona's r e s u l t s. He expressed doubt t h a t the selectivity of c o m-plex tissues could be achieved by o n e linking factor of the type that gives aggregates in tissue culture. It may be that Aaron is right and Marty is also right. Aaron's factors, whatever they turn out to be, may be only one kind of coarse adjuster. Certainly he would not claim that within the complexity of the retinal layer, all of the different cell interactions are being modulated by that factor. That is certainly not the case. So again, I will predict a hierarchy of specificities in cell-cell interaction.

So much for that. It is obvious I could get into specific details and diagrams. I hope that what I say provokes those of you who are interested in the connection of histocompatibility with a reasonable biological function. I think it is no task at all to explain polymorphism on this basis, because you are going to need a lot of these hierarchies, and it is going to emerge out of them. An example that Fritz and I discussed last night, is two separate genes linked cis with different products. If there is an advantage for these two to be expressed together for some functional reason, then they will stay together right on the chromosome. And if there is an advantage in a diversity of such genes, it will obviously be maintained. H2 tests may pick up the equivalent of allotypes, but not of idiotypes.

I have discussed a great number of structural things and a great number of functional things painting with a very broad brush. I think that, in spite of the incredible difficulties in the analysis of the surface membrane complex, the prospects are extraordinarily good. We are entering a new period of fractionation, microchemistry and better assays. By better, I mean more directly related to chemical binding. We also now have exquisite chemical methods to detect intermolecular interactions. Just to provoke Hardin, which I do all the time, out of admiration, I would say that I think the importance of what he said at this meeting related less to the angelic features of the halos around glycophorin, than it does to the fact that it is now possible to have a general method which will detect within these complex membranes specific intermolecular collisions and interactions and their specificity. That is no mean thing, and I believe the combination of all these methods is going to be really enormously powerful. Whether it will be suitable, of course, is another story, and I don't think that this will proceed smoothly. I think it is at this point where the medical people and those of you who are typing and doing all of this rather abstract serological work, are going to contribute by detecting structurally valuable accidents of nature.

It seems to me that our hopes and success should not be based on trying relentlessly to analyze the entire system. Rather, we should combine this analysis with accidents of disease involving deletions or additions to this kind of surface assembly. Medicine has shown itself to be capable of providing such examples time and again; a good case is Bence Jones protein and another is sickle cell hemoglobin.

We have come from the Iron Age of ferritin labeling through the Ice Age of freeze fracture and now we hope to arrive at the Golden Age in which we shall do all these marvelous things.

I can only exhort you not to be like the man who bought a cello and went home to his country estate and played on a single note for two weeks steadily, night in and night out, until his wife, losing patience, said to him, "I'm going out of my mind. Everytime I go to the city, I notice these guys moving their hands all over the fingerboard, and you are just sitting there like an idiot with your eyes rolled up playing the same note."

He said, "Those damn fools are looking for the note -- I have found it."

(Laughter)

Now, I invite your comments, some of which shall have to be out of order, because certain participants are leaving for airplanes. I think that is the fair way and I hope you do not mind.

MOSCONA: I want to say that although Jerry refers to his idea of motivation of cell surface as somewhat scandalous, in fact, there is quite a bit of evidence which suggests that it might not be a very outlandish idea. It is a fact that the surface properties of embryonic cells are modulated by contact with other cells, and this sort of system could provide an experimental approach to test the idea. In other words, there are experimental systems in which this idea could be tangibly explored and not left at a purely theoretical level.

EDELMAN: You remind me I had left something out of my Cassandra or anti-Cassandra predictions, the idea of a reconstruction experiment. It seems to me that even at levels more complicated than the kind Jackie and Hardin work at, we are going to be able to do some perturbation and reconstruction experiments by taking surface molecules and putting them into membranes to test just such interactions.

BACH: I think that the idea you bring up, Jerry, regarding the possible function of the major histocompatibility complex in differentiation is new. I think you are being perhaps generous and perhaps foolhardy to give credit to the ideas that have been brought up in the past. It is all related to the fact that within the one organism, there is presumed to be essential homogeneity of these cell surface markers. Their role in differentiation has been postulated in very largely abstract terms. Perhaps the most concrete of them is that the markers are expressed in different cell types and in different quantitative levels. Therefore, we know that in certain systems, although the same molecules may be involved, the quantities on the cell surface are important in differentiation.

What you are suggesting intrigues me, at least, because it says that within the organism, there can be variability superimposed. I think you turn to those of us using cellular system and say, well, you can test this by looking very carefully. I would turn around to you and say that those of you doing amino acid sequences on these molecules are much more likely to get an unequivocal answer than are we, although I am willing to go on looking. Let me just comment very briefly about the polymorphism.

EDELMAN: I wish you would, because I think that is an issue that is difficult.

BACH: The maintenance of polymorphism, as you know better than I, is a very difficult issue to explain because whereas ten years ago I think the general dogma has been that you had to have a selective advantage in order to maintain the polymorphism. The findings in Chicago with Drosophila locus having 6 to 8 alleles seems to be quite modest compared to the polymorphic systems we are now discussing.

It is not surprising that with serological tools of a very high discriminative capacity, you detect vast polymorphism, but we would still like to say there is a selective advantage based upon a unique supramolecular organization. Obviously, the products of the histocompatibility loci might function in the way you proposed in development. There may be many different alleles of these genes, carrying information which will increase diversity. I think, therefore, it could be, as you said yesterday, the driving force or the selective advantage toward maintaining polymorphism. I think one can extend this to linkage disequilibrium in quite straightforward fashion, if any of this is correct.

EDELMAN: Thank you. I do think the problem can be turned around another way. If you take minor loci, you may also be specifying surface molecules. I don't know if that is in fact proven. If you do assume that, and you take the position that what I have proposed is not the case, then every cell would contain an example of each antigen during development. But what I would, in fact, suggest is that it is not just a question of dosage, but it is a question of temporal sequence of expression as well as exclusion, involving a great variety of loci. There is no implication, however, in this idea that there is some kind of code in the membrane working like a combination lock.

I have carefully avoided, you notice, a detailed molecular model of recognition. I do that out of the last vestiage of caution that I have left. It is quite obvious that it is not going to be easy out of the blue to snatch an idea of what is being recognized. I gather from your statement, Fritz, that there is no evidence whatsoever that the system is a complementary system. I think that is important to rule this out right away: that in your mixed lymphocyte culture, the cell that is treated with mitomycin is using as those structures that are being recognized, the same

kinds of molecules as those which are recognizing it in the responding cell.
Therefore, complementation models of recognition, I think, are less likely.
I would certainly like you to comment on that.

It seems to me the key advance in this field is going to be
when someone devises an assay that measures binding, even relative binding, not just responses to binding. This has confused us all in immunology many times.

BACH: That has been brought up before yesterday afternoon. I don't
think that we can rule out complementarity at the moment, but I think
that there are experimental results which argue against it, but I would not
want to rule it out at this point.

Certainly, some of the minor histocompatibility antigens are
expressed on the cell surface. As a matter of fact, sitting in this audi-
ence, Charlie McKhann was one of the first to demonstrate that in order
to get antisera for these determinants, you need a concomitant H-2 difference. That brings up the question you just mentioned, that there is more
beta-2 on the surface than you can account for in association with HL-A.
Maybe this is going to be related not only to TL, as Jonathan Uhr dropped
as a pearl yesterday, but also to other histocompatibility antigens.

The last point I would want to make, as I think you quite
rightly said, we have different functions for various components of the
major histocompatibility complex. I tried to draw the function, given our
present limited armamentarium of testing the LD and SD-type of antigens,
but I tend to think we must get to the molecular level in assessing the gene
products using the tests.

EDELMAN: That is, the map of these functions as operationally proposed
is not necessarily the true functional map, but it will at least be reflective
of differences of function. I appreciate that. Now I think we are open to
discussion.

KAHAN: I think one important observation that did not come out in the
discussion, but is perfectly consistent with the hypothesis is that malignant cells seem to show expression of HL-A antigens which were not originally detectable in the primary host. This suggests that much of the
genetic information coding for histocompatibility factors is present in
every cell. The phenotypic expression may be determined by other factors, which might then be unmasked with malignant transformation.

EDELMAN: Does anyone wish to discuss that further? I don't want to
comment on it directly, but it seems to me an important point. The super-
vention of the malignant change or transformation can be a random and

unconnected event, that brings out the expression of a single gene product in a way that can be very useful -- myeloma is a very good example Everybody in the old days would say, "What are you studying myeloma proteins for? Everybody knows that they are cancer proteins and are abnormal." That seems funny now, but it is astonishing how many immunologists absolutely did not accept that these proteins were immunoglobulins. I think if you ask why, it was because of this analogous kind of reasoning, and also because the whole of immunology at that time was dominated by models of interaction that were not selective. Therefore, myeloma protein did not have any "activity" and if it didn't have any activity, it was not an antibody. Now I am turning over the discussion of any of these subjects or subjects that are your own.

McCONNELL: In a quantum mechanical system, one has a high sensitivity of degeneracy, that is a degenerate system of perturbation. As I understand it, your idea of the two sibling mice that start off genetically identical might end up with different cell surface components. Couldn't one conceive a real experiment in which the system is perturbed chemically or otherwise, so as to amplify those differences?

EDELMAN: Yes. Let me start with exactly the analogy you have brought out. Quantum mechanical degeneracy is represented by a splitting of energy levels or by alternative solutions to the Schrodinger equation and in fact is very sensitive to perturbation. But in fact the kind of degeneracy I have discussed may not be. For example, immunological degeneracy is so great that it is not a question of two or three levels. There are so many cells with different antibodies that can bind to a given antigen, that it is difficult to perturb the system except in an all or none fashion. You have put your finger on what I think is the key issue here. One should not go too far in either kind of analogy. The system of cell-cell recognition may not require as much degeneracy as the immune system but still may not be able to work if there are only a few alternative possibilities for recognition. It may lie somewhere in between. I think certainly we ought to be able to test whether perturbation would be useful. For example, I think degeneracy in the immune system shows the existence of sometimes hundreds of antibodies against a single antigen. In the response, however, there is the expression of only those of higher affinity through some kind of selective filter involving triggering. You can perturb such a system if you get at those high affinity cells. In the same way here, there may be cells at certain early stages of development that you could perturb in a fashion that would be absolutely critical. The effect of such perturbations would be different in the different models, that is in a degenerate and non-degenerate system.

　　　　Pedro, I don't want to perturb <u>you</u> -- but I think you brought

up a new idea at this meeting, and that was the notion that the c h o l e r a
toxin, or some portion of it, hooks onto one of the glycoplipids, a n d t h e
complex then somehow inserting itself into the membrane. The question
I directed to you and Dr. Hakamori was really two-fold. I realize on reflection I was not clear -- there are two kinds of competition experiments,
the "McConnell type" and the "Cuatrecasas type." The McConnell type is
non-specific competition in the sense of loading the membrane by inserting glycolipids. The Cuatrecasas type would be, I think, specific competition for binding to a receptor -- you see what I mean. Has anyone tried
just to insert a variety of glycolipids into a membrane over and above the
number that is already there? Do you have any ideas about this?

CUATRECASAS: I think it probably depends on the cell type. This is the
problem. In erythrocytes, it is very difficult to insert a large number of
glycolipids while, on the other hand, in other cells it is much easier t o
enter a large number. But the difference might be that the red cell has a
much larger number to begin with. This would suggest that there is a maximal capacity, at least for the kind fairly specific insertion, of t h e s e
glycolipids. I don't think I can really answer your question. I think i t
can certainly be tested chemically.

EDELMAN: That has not been done?

CUATRECASAS: No, It has not been done. It looks like there is a maximal capacity. One can examine insolubilized preparations of membranes,
molecules, and macromolecules, all of which combine fairly selectively
with the glycolipids. They all have a maximal capacity with given glycolipids. Whether there is compeition, cross reactivity, or cross competition is unknown.

EDELMAN: Have people tried interaction experiments with glycolipids in
vitro? Have they taken a whole bunch of them and added them to Steck and
Marchesi proteins to see if one sticks to this or that?

CUATRECASAS: It has not been done really. Your comments are a l s o
relevant in terms of the specificity of the carbohydrate. I think the situation is a little bit unusual with the cholera toxin recognizing the ganglioside. We would call the ganglioside the receptor, but in fact it is r e a l l y
two cholera toxin molecules, which are macromolecular proteins which recognizes the glycolipid. I think probably if this is a model of something
in biology relevant to glycolipids, it is really the toxin which is the receptor and the glycolipid which is the ligand.

EDELMAN: Yes. Marty, do you have any comments about the business of cell motion? How does it strike you? I think that if it were true that we had a submembranous assembly, that local movement, morphogenetic movement, and translocational movement could all severely affect n o t only the redistribution of receptors but also the whole state of the system -- don't you think that is a possibility?

RAFF: The simple answer is yes, and I think there are a number of examples that one could cite. I think there are difficulties in trying to generalize from any one system. For example, when one uses the lymphocyte as a model, one tends to neglect the fact that most cells in the body a r e interacting in a stable way with other cells, and there will certainly b e localized areas that are highly specialized and very different from other regions of membrane.

I wo u l d like to raise one other point. You tended to emphasize cooperativity in the interaction of cytoplasmic components w i t h membrane molecules; but I think there are many examples that suggest that some of these interactions can be remarkably localized -- for example the budding of viruses, phagocytosis and pinocytosis. Another example of localized and independent behavior of one region of a single cell and its membrane is that pointed out to me by Michael Bach yesterday. If antigen is applied locally to one region of a mast cell, it induces local degranulation in that region.

EDELMAN: I am grateful to you. I didn't mean to intend to exclude t h a t, which must take place. Certainly, the idea of cooperativity does not necessarily involve just positive cooperativity, nor does it have to be universal. It just seems to me that the structure is so complex that you can have all varieties.

Bernie, what Marty raised is a very important issue that I did not touch upon. Once you make a complex parenchymal organ by specific recognition, you may not need specific surface molecule to stabilize the structure. You get into the question of when structures of the type you have described play a role. The question is whether those structures are mainly architectural and relatively non-specific.

Has anyone tried a reconstruction experiment? I d o n' t remember if you mentioned w h e t h e r gap junction plaques from one tissue have been inserted in the cells of another. What I have in mind i s t h a t tissues may first establish specific connections and then differentiate structures like gap junctions. On the other hand, it may be that both k i n d s of structures are required at the same time.

GILULA: Developmentally, it has been observed in all the embryonic tissues that have been studied that junctions of communication arise q u i t e early in embryo. Concomitant with the origin of the junction of communication and with development, you have a severance of communication between organ systems.

In terms of the interaction between cells in a single organ, or cells of different organs, it is clear that in a culture system, you can generate interaction if an artificial type between cells from quite different types or tissues, so that the kinds of specificity of interaction to form these junctions are not clear at the present time. In fact, there seems to be a universal kind of interaction that can be generated in such an artificial system, and it is quite perplexing right now whether this interaction occurs within embryos and whether it does have a recognition purpose, that is, if there are recognition functions for communication. I don't know if that answers the question.

EDELMAN: Practically though, can you take your kind of structure and imbed them in other membranes of other cells? Can you actually transfer back into a real membrane. I am not just speaking of an artificial membrane.

GILULA: This has not been done yet, to my knowledge. Dr. McConnell and I discussed artificial reconstitution or recombinations. It is clear that if you take mutant cells and hybridize them with normal cells, the resultant hybrid has the ability to express the plaque, that is, the expression of the plaque seems to be dominant in all cases that have been observed. However, if you do an interesting experiment where you take intact tissue and dissociate it, the junction comes off on one cell of the complex. One cell of the two-cell association is void of the plaque, and the other cell has the entire plaque. No one has the answer yet to the question: when you put a cell with a plaque together with a cell without the plaque, does it use the entire plaque that came from the dissociation process. But it is possible to label the components, and see whether or not the plaque is reutilized by a cell system which has the capacity to generate it but does not have it performed.

EDELMAN: You had a question, Morris.

KARNOVSKY: It seems to me in relating to the question of movement that although we know that complexes and carbon particles and so on will slide around, I don't think anybody really knows if non-crosslinked components of the membrane are compartmentalized during movement and what control does occur. I think we concentrate on cells like lymphocytes, macrophages fibroblasts, and I think it would be very interesting to look at epithelial cells. For instance, nobody has shown capping can occur on a liver cell. It seems to me conceivably that the constraints of movement for membrane components in epithelial cells might be much more rigid than in the lymphocyte.

McKHANN: I would like to ask you a little more again on the biological value of polymorphism in the general sense. One can understand that a wide range of polymorphism is the basis of the selection for evolution, and is, of course, thereby of great value. On the other hand, polymorphism at an individual cellular level, particularly one that can be continuously renewed and expressed throughout the life time, may have some intrinsic dangers. One can visualize a situation where a change that gives a selective advantage to a particular cell type might be bad for the overall individual. So I would raise the question of whether polymorphism as expressed in the individual is really just a byproduct of that which is needed for evolution, or can you visualize something of intrinsic value to the individual?

EDELMAN: I want to stick to a very classical definition. I would not use the term "polymorphism" for variants within individuals that would arise somatically. Polymorphism refers to the expression of a diversity of allelic forms in the population. The diversity I was talking about is not the same kind of a thing. For example, it is not strictly correct to call variants of multiple genes for antibodies polymorphic unless there is evidence for allelism and segregation. If you accept that definition, my answer to your question is that somatic variation could be a dangerous situation if you allow the thing to go unbridled. Nonetheless, evolutionary mechanisms could easily be developed to prevent such a situation, as for example in tolerance. To control antibody levels, there must have developed systems to dampen the florid expression of the immune state. I think your question is well taken, but I don't think it poses a fundamental difficulty to these ideas, if you see what I mean.

POLLARD: What we've really talking about now, looking at the questions here on the board, is a sort of membrane mechanics, and implicit in the concept of membrane mechanics is energetics. In virtually all the things we are talking about, energy coupling exists and is specifically involved in capping. It is very clear that this depends on a mechanism dependent on energy coupling. I think that if we really want to get a chemical basis for any of the things you are talking about, then an understanding of energy coupling to this membrane mechanics is going to be important. Really, hardly anybody talks about it, and for good reason, we don't really know how that might happen, but it is an important idea.

EDELMAN: Yes, and one clear case is the Mitchell hypothesis in a very specified situation. Hypotheses of that type at least make that attempt. We have not yet verged on ideas like that here I guess. Maybe Hardin has thought about it, but I certainly know of no general discussion in terms you raised.

POLLARD: With regard to the Mitchell hypothesis, it would be quite possible to try to modify many of these membrane mechanical events by changing proton fluxes. You can do that experimentally.

EDELMAN: If someone can explain to me how when you take a cell reversibly to pH5 and back, you stop receptor mobility, I would be very glad. Of course, this doesn't relate directly to the Mitchell hypothesis, but you reminded me of it.

POLLARD: You need not change the pH by that much to test the Mitchell hypothesis.

EDELMAN: What is interesting in terms of the structure are the kinds of couplings, that is to say, driving forces, that we have not really discussed.

McCONNELL: I think I may have some ideas about this. This is relevant and I just don't know experimentally how you would do it. I did mention to you last night that there is the question of thermodynamic efficiency. In the saline system in which proton gradients are created in one place and utilized somewhere else, the lateral motion could increase the thermodynamic efficiency and serve as a driving force to bring components closer together.

POLLARD: I just want to say one thing -- it is all very well to talk about thermodynamic efficiency, and I think those things underlie all of the interesting cellular things that happen. But, however, one of the other things that cells may control is rates of phase separations and, perhaps, such modulations can be controlled at the energetic level. We don't understand this at all. On the basis of looking at liposomes, you might assume that a lot of things in membranes might happen quite spontaneously, and when you look at cells, they don't.

EDELMAN: Are there any other comments, general or specific? I think I will close this meeting which has been a rich one and has covered a lot of different subjects. We ought to express our thanks to the organizers, Barry Kahan, Ralph Reisfeld and our hosts here for their efforts to make us comfortable. We should be grateful to Mr. Wallach, the stenotypist, and his loyal cohorts for listening to our phonemes. Thank you all very much. I am going to close the meeting.
                    (Applause)   (The meeting was closed)

CONFEREES

Fritz H. Bach, University of Wisconsin, Madison, Wisconsin.

Daniel Branton, Harvard University, Cambridge, Massachusetts.

Pedro Cuatrecasas, Johns Hopkins University, Baltimore, Maryland.

Bruce A. Cunningham, The Rockefeller University, New York.

Gary S. David, Scripps Clinic and Research Foundation, La Jolla, Cal.

Gerald M. Edelman, The Rockefeller University, New York, New York.

Soldano Ferrone, Scripps Clinic and Research Foundation, La Jolla, Cal.

Heinz Furthmayr, Yale University, New Haven, Connecticut.

Norton B. Gilula, The Rockefeller University, New York, New York.

Victor Ginsburg, National Institute of Health, Bethesda, Maryland.

Sin-itiroh Hakomori, University of Washington, Seattle, Washington.

Roger W. Jeanloz, Harvard Medical School, Boston, Massachusetts.

Barry D. Kahan, Northwestern University Medical School, Chicago, Illinois.

Morris J. Karnovsky, Harvard Medical School, Boston, Massachusetts.

Howard M. Katzen, Merck Institute for Therapeutic Research, Rahway, N.J.

Stephen J. Kennel, Scripps Clinic and Research Foundation, La Jolla, Cal.

Richard Lerner, Scripps Clinic and Research Foundation, La Jolla, Cal.

Michael C. Lin, National Institute of Health, Bethesda, Maryland.

Vincent T. Marchesi, Yale University Medical School, New Haven, Conn.

Harden M. McConnell, Stanford University, Stanford, California.

Charles F. McKhann, University of Minnesota, Minneapolis, Minnesota.

Aron A. Moscona, University of Chicago, Chicago, Illinois.

Neal R. Pellis, Northwestern University Medical School, Chicago, Illinois.

Jack H. Pincus, Northwestern University Medical School, Chicago, Illinois.

Harvey B. Pollard, National Institute of Health, Bethesda, Maryland.

Miroslav D. Poulik, Wayne State University, Detroit, Michigan.

Gerald B. Price, Ontario Cancer Institute, Toronto, Ontario, Canada.

Martin Raff, University College of London, London, England

David F. Ranney, Northwestern University Medical School, Chicago, Illinois.

Ralph A. Reisfeld, Scripps Clinic and Research Foundation, La Jolla, Cal.

Jacqueline Reynolds, Duke University Medical Center, Durham, N.C.

Sandra Ristow, University of Minnesota, Minneapolis, Minnesota.

Martin Sonenberg, Memorial Sloan-Kettering Cancer Center, New York.

Gnanasigamoni Sundharadas, University of Wisconsin, Madison, Wisconsin.

Theodore L. Steck, University of Chicago, Chicago, Illinois.

Motowo Tomita, Yale University, New Haven, Connecticut.

Gordon M. Tomkins, University of California, San Francisco, California.

Jonathan W. Uhr, University of Texas, Dallas, Texas.

Peter Wernet, The Rockefeller University, New York, New York.

Ichiro Yahara, The Rockefeller University, New York, New York.

Page 274

1. B. D. Kahan
2. G. Tompkins, G.M. Edelman, N.B. Gilula
3. A. A. Moscona
4. N. B. Gilula
5. Conference Session
6. V. T. Marchesi
7. M. Raff, T. Steck, H. Pollard
8. M. Lin, T. Steck, P. Cuatrecasas, H. Katzen

Page 275

9. M. Karnovsky
10. S. Hakamori, V. Ginsberg
11. R. A. Reisfeld
12. E. Appella
13. H. McConnell, D. Branton, F. Bach
14. R. Jeanloz
15. R. Lerner, N. Pellis, S. Kennell, R.A. Reisfeld
16. G. M. Edelman, S. Hakamori
17. J. Reynolds, S. Ristow, C.F. McKhann

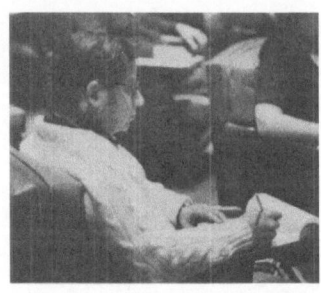

1

2

3

4

5

6

7

8

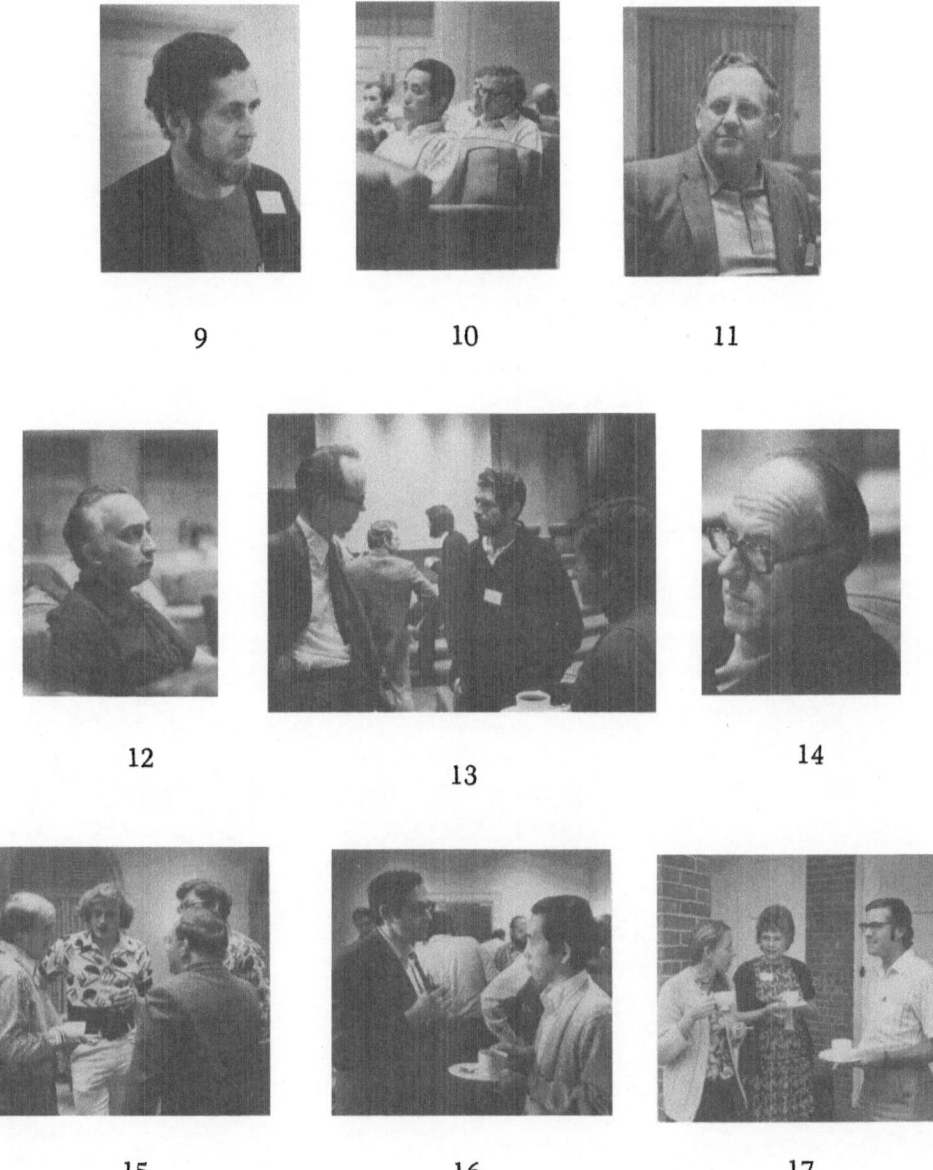

9   10   11

12   13   14

15   16   17

Adrenal medulla
 chromaffin granules, 65-67
 exocytosis and membrane fusion,
  65-67
 ultrastructure, 66
β_2 Microglobulin
 amino acid sequence, 138
 association with histocompatibility
  antigens, 140
 association with TL antigens, 178
 homology with Ig structure, 139-140
 in body fluids, serum, 94
       urine, 94
 specific antibody, 96-97
  amino acid sequence, 94, 138
  association with HL-A antigens,
   95-96
  distribution, 95
  effect on MLC reaction, 97-98
  homology with immunoglobulin,
   94
  molecular weight, 94
  utilization for HL-A isolation,
   97
Carbohydrates
 aggregation between carbohydrate
  molecules, 12-13
 carbohydrate-carbohydrate interactions, 11-13; 56-57
 tumor cells, 59
Cell-mediated lympholysis, 170
Cell-recognition
 embryonic vs. immunologic, 155
Cell surface carbohydrates, 51-60
 assessment by antibody precipitation, 52-54
 chemical structures, 51-54
 detection by lectins, 55-56
 detection by radioimmunoassay, 54-
  55

Cell surface carbohydrates
 free oligosaccharides in human
  milk, 51-52
Cholera toxin
 binding to gangliosides, 43-45
 subunit structure, 43
Choral hydrate, solvent for membranes, 176-177
Complement-mediated cytotoxicity
 binding of HL-A antibody, 188
 complement component binding, 188
 effect of complement activation
  pathway, 189
 susceptibility during cell growth
  cycle, 187-192
Concanavalin A
 activity, assessed by agglutination,
  82
 capping of receptors, 122
 carbohydrate composition, 81
 difference from Ig capping, 122-123
 effect on cap formation, 116
 effect on DNA synthesis, 143-145
 effect on Ig receptor distribution,
  115-117
 isolation from membranes, 81
 mechanism of mitogenesis, 140-141
 molecular weights of, 81
 receptors on lymphocytes, 80-82
 regulation of receptor mobility, 118
 solubilization of, 80-81
 stimulation of spleen cells, 141-142
 suppression of inhibitory effect, 117
Dielaidoyl phosphatid choline (DEL),
  103
 freeze fracture, 103-104
 spin label, 103
Embryonic cell recognition, 155-157
Erythrocyte ghosts
 blebbing and aggregation, 19-21

Erythrocyte ghosts
  distribution of membrane particles,
    17-21
  freeze-fracture, 17
  protein-protein interactions, 17-18
Erythrocyte membrane
  extraction with LIS, 5-6
  fibrillar proteins and penetrating
    glycoproteins, 22
  major polypeptides in, 5-8
  particle heterogeneity, 27
  PAS, and PAS$_2$, 6
  protein components, 4-16
  protein-glycoprotein interactions,
    22-25
  SDS-polyacrylamide gel patterns,
    5-6
  spectrin in, 5
Gap junctions
  electrophoretic profile of proteins,
    135
  rat liver membranes, 131-132
  ultrastructure, 132-134
Globosides, incorporation into mem-
    branes, 46
Glycolipids
  analysis by antibody digestion, 38-
    39
  binding with cholera toxin, 40-45
  biological function, 39-40
  effect of sialic acid, 41
  glycosphingolipids, chemistry, 37-
    38
  labeling with galactose oxidase, 39
  localization in cells, 38-39
  membrane components, 37-48
  organization within membrane, 39
Glycophorin
  effect of bound lipid on aggregation,
    15-16
  hydrophobic portion conformation,
    14-15
  isolation, 6-8
  lateral membrane distribution, 112

Glycophorin
  phosphorylation, 13-14
  proteolytic peptides, 8
  spin label probes of, 106
  structural features, 6-7, 13-15
Glycoproteins
  behavior in SDS, 8-14
  binding of SDS, 9
  H-2 antigenicity of, 61
  isolation from tumor cells, 57-60
  limitations of molecular size esti-
    mation, 8-11
  micelle size in SDS, 9
  molecular size, 59
  nature of SDS micelle, 10-11
  reaction with lectin of V. grami-
    nea, 59
H-2 antigens
  accessibility to ligands, 31
  carbohydrates of, 199-200
  cellular and humoral tolerance
    induction, 195-197
  distribution on lymphocytes, 30-34
  effect on tolerance in neonates,
    197-199
  expression on different cell types,
    32-33
  histocompatibility complex,
    genetics, 167-168
    Ir-IA and Ir-IB loci, 168
    LD loci of, 167-168
    SD loci of, 167-168
  solubilization by papain, 194-195
HL-A antigens
  assay of soluble antigens, 190-192
  association with β_2-microglobulin,
    89, 92
  attachment to membrane, 33
  carbohydrate presence, 199-201
  complex, LA, FOUR, and AJ loci,
    168-169
  disposition, 36
  expression during cell cycle, 87-89
  mechanism, 83, 87

HL-A antigens
  molecular nature of, 89, 92
  penetration, 32-33
  radioiodination profile of, 91
  solubilization with KCl, 82-86
  soluble HL-A antigen in serum, 193
  surface expression on lymphocytes,
    86-93
Immunoglobulin
  cap formation,
    cytochalasin B effect, 114
    microfilaments, 115
  measurement, 30-31
  on lymphocytes, 30-31
  patch formation, 114
  radiolabeling, 30-31
  receptors, 114
Insulin receptors
  affinity chromatography, 224-225
  binding of insulin, 221-228
  chemical nature, 229
  insulin-degrading enzyme, 224-227
  irreversibility of receptor: insulin
    complex, 227-228
  solubilization, 221-222
Ionic coupling between cells, 130-136
  distance between plasma mem-
    branes, 131
Lactoperoxidase catalyzed radio-
    iodination of cell surfaces,
    62-64
LD antigens and loci, 167-175
  in man, 168
  in murine system, 167
  in other species, 169
  response in MLC, 167
  role in mouse allografts, 171
Lymphocyte membrane
  cap formation of, 114-115
  macromolecules of, 30-37
  modulation by cellular structures,
    113
  patch formation of, 114-115
  receptors, relation to intramembra-
    nous particles, 123

Liver membranes,
  fluidity, 108-109
  fluorescence polarization, 109
Major Histocompatibility Complex
    (MHC)
  genetics of H-2 complex, 167-168
  genetics of HL-A complex, 168-169
Membranes
  dielaidoyl phosphatidylcholine
    (DEL), 103
  dipalmityl-phosphatictyl choline
    (DPL), 103
  glycoproteins, 3-8
  lateral interactions, 103
  lateral phase separation of lipids,
    104
  perpendicular transport through,
    103
Metabolic cooperation between cells,
    130-136
Mixed leukocyte reaction and, 83-84
  mechanism of action, 84-85
Mixed lymphocyte reaction (MLC)
  determinants of, SDS gel profile,
    179-180
  inhibition by human alloantisera,
    179
  LD products, 170
  primary targets, 170
Monosialoganglioside (GMI)
  binding to cholera toxin,
  molecular nature of, 42
  receptor for cholera toxin, 42
Murine Ia antigens
  cell distribution of, 177
  molecular size of, 177
Non-specific Immunosuppression
    in malignancy, 206-207
Plasma membranes
  adrenal cortical cells, 72-73
  adrenal medulla, 64-72
  glycoprotein electrophoretic profile,
    69-73
  glycoprotein orientation and location,
    74-76

Plasma membranes - continued
  isolation of, 72
  protein components, electrophore-
    tic patterns, 67-69
  reconstitution of, 73
  secretory vesicle association with,
    77-79
  soluble proteins, 72
  structure and fusion, 64-65
  ultrastructure of isolated mem-
    branes, 69
  vesicle complex, 73-74
Polymorphonuclear leukocytes
  Ig capping inhibition with Con A,
    124-125
Receptor-steroid complex
  activation of, 148
  amount in cytoplasm, 149
  effect on intracellular behavior,
    147-155
Red cell ghosts
  blebbing, 19
  freeze-fracture analysis, 17-21
  intramembrane profiles, 18
  spectrin, 18-20
SD antigens and loci, 167-175
  in man, 169
  in the murine system, 167
  in other species, 169
  response in MLC, 167
  role in mouse allografts, 171
Serum blocking factors
  effect on tumor growth, 201-202
Sialic acid on tumor cells, 58
  action of proteases, 58
  density on surface, 58
  removal with neuraminidase, 58
Soluble tumor antigens
  assay, in vitro, 216-217
  immunogenicity, in vivo, 211-216
  parameters of immunogenicity,
    in vivo, 214
  purification of, 217-218
  solubilization, 211-212, 216

Soluble tumor antigens - continued
  suppression of lymphocyte respon-
    siveness, 202-206
Spectrin
  effect on red cell stability, 24
  location on membrane, 28-30